1 MONTH OF
FREE
READING

at
www.ForgottenBooks.com

By purchasing this book you are eligible for one month membership to ForgottenBooks.com, giving you unlimited access to our entire collection of over 1,000,000 titles via our web site and mobile apps.

To claim your free month visit:
www.forgottenbooks.com/free909953

ISBN 978-0-266-92049-6
PIBN 10909953

AD_____
T.M. 40

Pleistocene-Holocene Sediments Interpreted by Seismic Refraction and Wash-Bore Sampling, Plum Island-Castle Neck, Massachusetts

by

Eugene G. Rhodes

TECHNICAL MEMORANDUM NO. 40

JULY 1973

U.S. ARMY, CORPS OF ENGINEERS

COASTAL ENGINEERING RESEARCH CENTER

Reprint or republication of any of this material shall give appropriate credit to the U.S. Army Coastal Engineering Research Center.

Limited free distribution within the United States of single copies of this publication has been made by this Center. Additional copies are available from:

National Technical Information Service
ATTN: Operations Division
5285 Port Royal Road
Springfield, Virginia 22151

At the time of publication, prices were $3.00 for hard copies and $0.95 for microfiche.

Contents of this report are not to be used for advertising, publication or promotional purposes. Citation of trade names does not constitute an official endorsement or approval of the use of such commercial products.

The findings of this report are not to be construed as an official Department of the Army position unless so designated by other authorized documents.

Pleistocene-Holocene Sediments Interpreted by Seismic Refraction and Wash-Bore Sampling, Plum Island-Castle Neck, Massachusetts

by

Eugene G. Rhodes

TECHNICAL MEMORANDUM NO. 40

JULY 1973

U.S. ARMY, CORPS OF ENGINEERS
COASTAL ENGINEERING
RESEARCH CENTER

ABSTRACT

The wash-bore method of soil sampling, which can be employed from both floating and land equipment, was found to be an excellent technique for subsurface study. Seismic refraction is a valuable tool in the coastal environment but phenomena to be considered when interpreting records include a) "the blind zone," b) the non-zero time intercept, c) time gaps in the time-distance plots over buried peat, and d) variable thicknesses of dry sand layers.

The seismic method successfully located Pleistocene and bedrock topography. Dry sand, water-saturated sandy sediments, glacial till, and bedrock provide well-defined seismic contrasts. Glaciomarine clay does not show a seismic contrast with respect to sandy, water-saturated sediments.

Topography exposed during lower sea level 10,000 to 11,000 B.P. (Kaye and Barghoorn, 1964) has a dominant influence on modern coastal geology. Barrier-island features became anchored on drumlins and other Pleistocene features as the relative sea level rose. Thick sequences of sediments, including clam-flat facies, channel deposits, and point-bar deposits, accumulated in the estuaries behind these barrier beaches. Major channels of the estuaries migrated landward with the sea-level rise, a hypothesis in agreement with McCormick (1968).

A bedrock high beneath the centerline of the barrier beaches on all seismic profiles is perpendicular to barrier elongation. Bedrock topography is extremely irregular and drops over 150 feet beneath the Parker and Essex estuaries. Essex Bay appears to move in deep channels incised in underlying bedrock; two mid-estuarine sand bodies, Essex flood-tidal delta and Middle Ground, appear to have rather stable sedimentary deposits.

Although no radiometric dates were determined for samples taken in this study, the sedimentary stratigraphy fits the time frame of McIntire and Morgan (1963) and Kaye and Barghoorn (1964).

FOREWORD

CERC is publishing this report because of its interest and value to coastal engineers.

This report was prepared by Eugene G. Rhodes under Contract No. DACW-72-70-C-0029 with CERC. It is an edited version of a thesis submitted in partial fulfillment for a M.S. Degree in Geology at the University of Massachusetts.

Special appreciation is extended to Miles O. Hayes, who served as the immediate supervisor of the study from inception to completion. Professors Joseph H. Hartshorn and Randolph W. Bromery provided a constructive,

iii

critical review of the geological and geophysical problem. The wash-boring phase of the study could never have been accomplished without the equipment and instruction from the Civil Engineering Department, University of Massachusetts. Professor Karl Hendrikson contributed to the project through the training and advice to the author and the field crew.

The author is especially grateful to George E. Butler, Jr. and Richard K. Callahan. In addition, John F. Kick, Stewart C. Farrell and Jon C. Boothroyd gave considerable effort to the field study.

Edward S. Moses, manager of the Parker River Wildlife Refuge, obtained approval for field operations on Plum Island. Robert Hebb, Department of Natural Resources, also offered his cooperation for the author's use of the Plum Island State Park. The Trustees of Reserva-tions, in particular Garret Van Wart and Gordon Abbot, Jr., opened the Richard T. Crane, Jr. Memorial Reservation, to both seismic and drill-hole phases of the study. Additional gratitude is expressed to the employees of the Parker River Wildlife Refuge and the Crane Reservation for their help during the project.

Vincent J. Murphy of Weston Geophysical, Inc. and Curtis R. Tuttle of the United States Geological Survey, aided greatly in the reduction and interpretation of seismic data. William G. McIntire and James M. Coleman of the Coastal Studies Institute, Louisiana State University helped with geological interpretations.

At the time of publication, Colonel James L. Trayers was Director of CERC; Thorndike Saville, Jr. was Technical Director.

NOTE: Comments on this publication are invited.

This report is published under authority of Public Law 166, 79th Congress, approved 31 July 1945, as supplemented by Public Law 172, 88th Congress, approved 7 November 1963.

CONTENTS

Page

I.   INTRODUCTION   i

II.   FIELD TECHNIQUES   4

      1. Drill-Hole Study . . . . ' 4
      2. Seismic Refraction Study 13

III. SEISMIC PROFILE INTERPRETATION 20

      1. Formulas and Nomograms 20
      2. Seismic Velocities 20
      3. Graphic Interpretation 22
      4. Problems of Interpretation 25

IV.   DRILL-HOLE INTERPRETATION. 41

      1. Generalized Stratigraphy 41
      2. Glacial Till and Glaciomarine Clay 41
      3. Indurated Clay 43
      4. Weathered Zone 45
      5. Black Peat 45
      6. Estuarine Sediments. 45

V.   CASE HISTORIES FROM PLUM ISLAND. 47

VI.   DEVELOPMENT OF A BARRIER ISLAND. 55

VII. CONCLUSIONS. 57

      LITERATURE CITED 59

      APPENDIX - SUMMARY LOGS OF DRILL HOLES 63

TABLE

Listing of Bore Holes Containing Marine Clay 44

FIGURES

1. Location Map for Study Area.   2

2. Location Map for Drill Holes and Seismic Lines   5

3. Drill Rig on Barge Platform.   6

Page

4. Drilling Crew Disconnecting Pull Piece   7

5. Cathead Acting as a Rope Clutch.   8

6. Diagram of Drill Rig   9

7. Washbit. 10

8. Two Common Samplers and Some Sample Retainers. 12

9. Geospace GT-2A Portable Refraction System. 15

10. Geophone in Soil 16

11. Type 47 Polaroid Film Record 17

12. Diagram of Seismic Profile Layout. 18

13. Offset Distance and Critical Angle 21

14. Apparent and True Bedrock Velocities 23

15. Determination of Bedrock Relief by AT. 24

16. Example of a Non-Zero Intercept. 26

17. Apparent Sediment Velocities Due to Variable
    Thickness. 27

18. Time-Distance Plot Showing a Time Step for Buried
    Peat 28

19. Profile from PI-7 and Drill Hole PIC 29

20. Time-Distance Plot and Cross Section from PI-17. 31

21. Time-Distance Plot from Middle Ground. 32

22. Cross Section of Middle Ground Showing
    a Bedrock High 33

23. Graphical Explanation of the "Blind Zone". 35

24. Nomogram for "Blind-Zone" Computation. 36

25. Transverse Profiles Showing a Bedrock Rise 38

FIGURES (continued)

Page

26. Cross Section from CB-18, CB-19 and Drill Hole CBG   39

27. Essex Flood-Tidal Delta Profile.   40

28. Sandy Till from Drill Hole CBB   42

29. Weathered-Zone Material.   46

30. Sketch of Castle Neck at 10,000 B.P.   48

31. Sketch of Southern Plum Island at 10,000 B.P.
    and 4,000 B.P.   50

32. Sea-Level Trends for Plum Island Area.   52

33. Sea-Level Curve.   53

34. Origin of Castle Neck.   56

PLEISTOCENE-HOLOCENE SEDIMENTS INTERPRETED
BY SEISMIC REFRACTION AND WASH-BORE SAMPLING,
PLUM ISLAND-CASTLE NECK, MASSACHUSETTS

by

Eugene G. Rhodes

I. INTRODUCTION

The shoreline of northeastern Massachusetts is dominated by extensive barrier beaches. This study attempts to reconstruct the three-
dimensional stratigraphic framework of these deposits. The Pleistocene
and Holocene stratigraphy of Castle Neck and the southern third of Plum
Island, the two largest barrier beaches in this area (Figure 1), was
determined by wash-bore and seismic refraction methods.

Thoreau vividly described this part of the New England coast in the
1840's:

> "It is a mere sandbar exposed, stretching nine miles parallel
> to the coast, and, exclusive of the marsh on the inside, rare-
> ly more than half a mile wide. ...The island for its whole
> length is scalloped into low hills, not more than twenty feet
> high, by the wind, and excepting a faint trail on the edge of
> the marsh, is as trackless as Sahara. ...I have walked down
> the whole length of its broad beach at low tide, at which time
> alone you can find a firm ground to walk on, and probably
> Massachusetts does not furnish a more grand and dreary walk.
> ...A solitary stake stuck up, or a sharper sand hill than
> usual, is remarkable as a landmark for miles; while for music
> you can hear only the ceaseless sound of the surf, and the
> dreary peep of the beach birds."
> from Henry David Thoreau, *A Week on the Concord and Merrimack
> Rivers.*

The 3-mile expanse of dunes and beach south of Plum Island, which
comprises Castle Neck, could also have been the object of Thoreau's
description. Much of Plum Island is now controlled by the Parker River
Wildlife Refuge. Castle Neck is a recreation area held by the Trustees
of Reservations, Milford, Mass. Both areas have been subjected to agri-
culture and other manmade modifications since colonial times. Yet, today
they are two of the least disturbed sand bodies on the New England coast.

Much of Plum Island and most of Castle Neck have been allowed to
undergo uninterrupted natural modifications in response to the environment. For this reason the area provides an excellent outdoor laboratory
for this type of project.

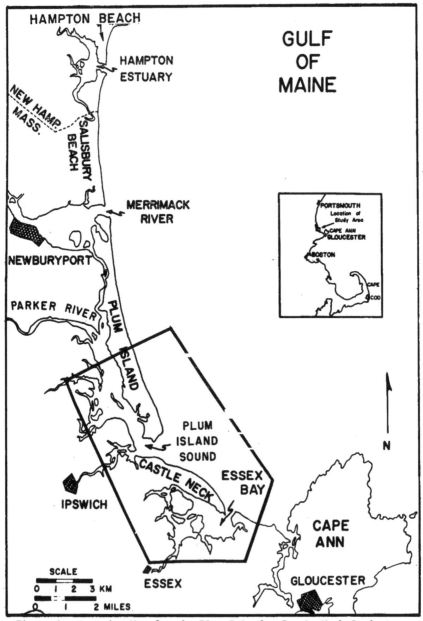

Figure 1. Location Map for the Plum Island - Castle Neck Study Area

Sears (1905) found the coastline to be subsiding at about 3 feet per century, and noted the various drift patterns of the beach sand. Chute and Nichols (1941) presented ideas on the formation of Castle Neck, and studied the recent geologic history of the area in greater detail than previous workers. Sammel (1963) mapped the surface geology of the Ipswich quadrangle, which includes parts of Plum Island and Castle Neck. McIntire and Morgan (1963) were the first to probe the subsurface with drill holes, and to correlate drill-hole horizons with known Pleistocene and Holocene stratigraphy.

From 1965 to 1972, students of Professor Miles O. Hayes, at the University of Massachusetts, studied coastal processes in this area. Much of this material remains unpublished in the form of theses. A large part of this work has been summarized in *Coastal Environments*, a guidebook for the Society of Economic Paleontologists and Mineralogists field trip, May 9-11, 1969, written by the Coastal Research Group, University of Massachusetts. McCormick (1968) studied marsh stratigraphy in the Plum Island marsh with more than 60 cores, some to a depth of 40 feet.

Linehan (1942, 1948) was one of the earliest seismologists to do shallow refraction work in New England. He and others used the method to determine bedrock profiles along proposed highway routes. By this method, the Department of Public Works determined the volume of overburden to be excavated and the nature of the rock topography. The seismic method has been used to locate preglacial bedrock channels of the Merrimack, Connecticut, and Charles Rivers. In these examples, refraction seismology was used primarily to "see through" the overburden and map the bedrock surface. Currier (1960) evaluated the seismic method for determining bedrock topography under proposed highway and foundation sites. He also discussed resistivity, wash-boring, and core drilling as useful subsurface profiling aids, but concluded that the seismic method was the most successful.

The purpose of this study was to develop field and interpretive techniques for the study of the third dimension in the coastal environment. Plum Island and Castle Neck were chosen for their almost untouched natural state and for the large amount of accumulated information on surface processes. An understanding of the types and rates of shore processes aids in the interpretation of subsurface data.

Methods and interpretative guidelines developed during this study can now be applied in other areas, perhaps while gathering other coastal data. The result can be a three-dimensional sedimentary model that effectively links space and time.

II. FIELD TECHNIQUES

1. Drill-Hole Study

Seventeen wash-bore holes ranging in depth from 30 to 100 feet were completed in this study (see Figure 2). This boring program employed an Acker Model RGT wash-bore rig operated by a three-man crew. Figure 3 shows the barge-mounted drill rig at an estuarine drill site. The maximum drilling depth with this equipment is 100 feet. Except where till or bedrock was encountered, the bore holes were drilled to that depth. Bore holes were logged continuously by noting the nature of the washwater. Holes were sampled at 5- to 10-foot intervals, or at every change of sediment type.

The wash-boring process consists of hammer driving a casing, washing the inside of the casing clean, and then sampling beyond the end of the casing. This project used an Acker NX flush-coupled casing, which has an inside diameter of 3 3/16 inches. It is constructed of cold-drawn, steel tubing and assembled in 5-foot lengths using couplings with square threads. It is designed to withstand only limited driving, and is not recommended for casing drill holes in sand beyond 100 feet.

The casing was driven into the sediment using a 300-pound drive hammer sliding on a pull piece attached to the casing, and was retrieved by bumping the hammer against the top of the pull piece. Figure 4' shows the drill-rig crew disconnecting the pull piece and a 5-foot section of casing during casing removal. Casing should be driven into the sediment vertically to eliminate sway in the hammer and prevent metal fatigue near the connecting joints.

The hammer was attached to a 1-inch manila line by a manila sling, rather than by a chain or cable sling, to protect the threads on top of the pull piece. The 1-inch manila line passed through a pulley on top of the drilling mast and down to the cathead on the power unit. Figure 5 shows the manila line engaging around the cathead as the operator applies tension to the slack end of the line. The manila line acts as a clutch, engaging on the cathead when tension is applied and releasing when the operator lets the line go slack. Casing is driven by alternately tightening and slackening the line to the cathead. Figure 6 shows the drill rig and its operation. The number of blows per foot of penetration is recorded to obtain useful penetration data. The casing is cleaned of sand by a chopping bit (Figure 7) before taking a sample.

A high-pressure water jet is used; there is no rotary movement. A working pressure of more than 100 pounds per square inch is developed by a double-action piston pump opposite the cathead. Water is supplied to the pump directly from a pond, estuary, or truck-mounted water tank. A day's drilling, which usually averages 50 feet, requires about 1,000 gallons of water. Saline, brackish or fresh water may be used if precautions are taken to retard rust after use.

4

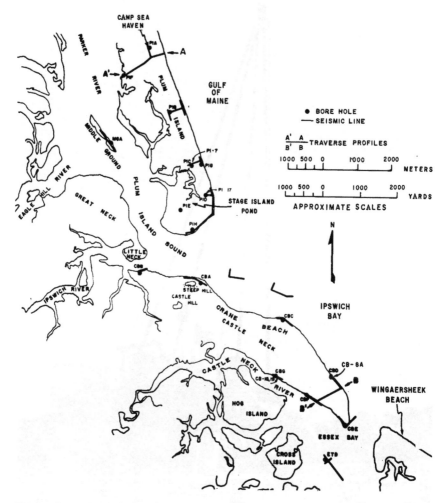

Figure 2.   Location Map for 17 Drill Holes and Important Seismic Lines.
Seismic line numbers mentioned in text are shown.   The Essex
flood-tidal delta site is at drill hole ETD (Figures 3 and 27).

Figure 3. Drill Rig on Barge Platform. This site is on the Essex
flood-tidal delta where drilling was done at low tide for
added stability. The casing is driven through a well in
the center of the barge.

6

Figure 4. Drilling Crew is Disconnecting the Pull Piece Before Removing
a 5-Foot Section of NX Casing. After removal, the pull piece
will be attached to the next 5-foot section and more casing
will be bumped out of the ground.

Figure 5. The 1-Inch Manila Line Engages Against the Cathead as a Rope Clutch When the Operator Applies Tension to the Free End. This unit has a 500-pound hammer capability, but can lift heavier weights.

8

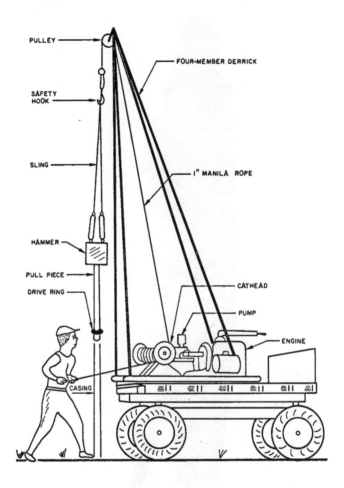

Figure 6. Diagram of the Wash-Bore Unit Used. A 15-horsepower engine powers a clutch unit which in turn powers the cathead and water pump. The pump, a double acting piston type, develops over 100 pounds per square inch. A high-pressure hose carries this water to the drill string. The operator uses the mechanical aid of the cathead to raise and lower heavy objects and to operate the hammer.

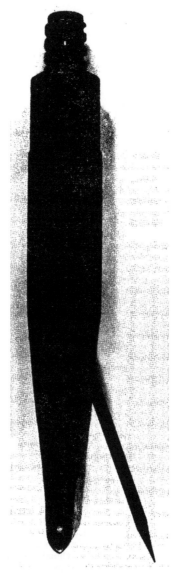

Figure 7. Washbit Used for Cleaning the Casing Before Sampling. This
is a chopping bit, and jets water from both sides. The
pencil locates one jet.

Water is pumped through the center of either E (1 5/16") or A (1 5/8") size drill rod by a water swivel on the upper end of the drill-rod string. This swivel allows rotation of the drill string in the hole without twisting the hose which carries the water from the pump to the drill-rod string.

When drilling in sandy material, wash water is no longer returned up the hole as the washbit passes beyond the end of the casing because of the porosity of the sand. The drill-rod string is removed from the hole in 10-foot sections, and a sampler is lowered into the hole by reassembling the drill string. A sample is taken beyond the end of the casing before advancing it 5 more feet.

Samplers are of several types. The type to use depends on the sediment being sampled and the preference of the operator. Figure 8 shows two more common samplers used in shallow holes. The split-spoon sampler is used in sand, clay or till; the Shelby tube works best in silt or clay.

A variety of sample retainers are used in the shoe of the split spoon. The more successful ones are also pictured in Figure 8. The metal spring fingers with the plastic bags surrounding the fingers proved most dependable. Prophylactics of latex rubber are far superior to the plastic bags sold as retainers, and are a recommended substitute. When working in clays with either the Shelby tube sampler or the split spoon, no retainer is necessary if the ball check valve in the head of the split spoon and the Shelby tube is kept free of sediment. The ball valve should be flushed clean with water before sampling to prevent sediment from being driven into it. Unless a heavy-duty, split-spoon sampler is used near gravel or till, metal fatigue caused by forced penetration of the sampler will result in a lost sampler.

The presence of clay or fine silt the entire length of a drill hole eliminates the need for casing. Clay is self-casing and retains the hole even if drilling is suspended overnight. Many crews use a drilling mud to case sandy parts of a hole, because mud mixed with water fills the pores of the sand and causes the hole to retain its shape. In very porous sand, this process may be tedious. Here, the choice between drilling mud or steel casing depends on the type of equipment available and the preference of the crew. The use of mud simplifies drilling deeper holes as there is no casing to be retrieved. However, certain pumps cannot handle the clay-water mixture.

A common problem results from sandy till or gravel under a thick layer of glaciomarine clay. Because the clay part of the hole is not cased, and could be cased only with great difficulty, there is a water loss when the washbit passes from the clay into the more porous material beneath. This usually blocks further drilling, and is one of the more difficult problems of using nonrotary, limited-capability equipment.

11

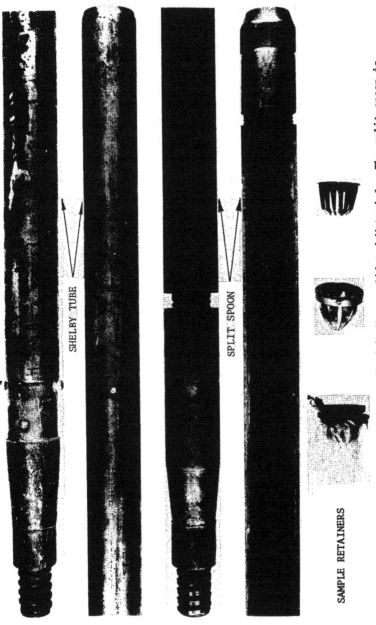

SHELBY TUBE

SPLIT SPOON

SAMPLE RETAINERS

Figure 8. Two Common Samplers Used in Unconsolidated Material. The split spoon is successful in all types of sediments; the Shelby tube is best in clay. At the bottom are various sample retainers placed in the split spoon when it is assembled. The spring bag covering is the most dependable.

An experienced crew can drill up to 50 feet per day. The early appearance of clay in a hole allows speedier progress. A crew can occasionally drill and sample a 100-foot hole in one long day. But usually, it takes 2 1/2 to 3 days to do a good job on a hole. This includes setup and takedown time for the rig and transportation from the previous site.

Estuarine (barge-mounted) operations are slower, but unlimited washing water (over the side of the barge) is an obvious advantage. Anchoring a rig against tidal currents is not too difficult. Drilling is then done through a well in the center of a barge with the casing extending up to deck level. This casing not only facilitates re-entry but also helps to anchor the barge. A site on an intertidal bar is desirable because drilling can be done at low tide while the drilling base is grounded.

2. Seismic Refraction Study

Seismic refraction principles were used in earthquake seismology before their application in shallow refraction surveys. The principles of refraction remain the same whether the layers being studied are thousands of feet thick or merely tens of feet thick. Refraction surveys enable the scientist to gather data on the geometry and composition of subsurface materials. The simplest case is a two-layer problem consisting of a high-speed layer overlain by a low-speed layer, such as crystalline bedrock overlain by glacial material.

In the simple two-layer case, a seismic wave is generated at the surface and travels away from its propagation point in all directions at a velocity unique to the upper layer. On crossing the interface with the lower layer, the wave moves at a higher velocity and overtakes the surface wave. Geophones placed in a linear fashion away from the point of energy propagation measure the arrival time of the direct (surface) and indirect (subsurface) waves. A time-distance plot of these arrival times will determine when the indirect wave overtakes the direct wave, allowing determination of the depth to the subsurface layer.

The field techniques of the shallow-refraction phase are similar to those of the drilling phase. In both phases, equipment must be transported to sites not easily accessible. Dependable four-wheel drive vehicles and rugged boats are needed. Seismic study requires the use of bulky, but delicate, rolls of wire and explosives; both can prevent transportation problems.

The seismic work was done with the Geospace Model GT-2A portable refraction system. This unit works well near the shore because it is weatherproof when the instrument cover is closed, and is rugged and dependable over a broad range of temperatures and humidities. All time records are taken on type 47 Polaroid film, allowing rapid review of seismic results in the field. Figure 9 shows the recorder with its Polaroid camera in position for recording. Figure 10 shows the geophone attached to its cable takeout. Phones are always planted in holes below the vegetation or loose soil layer. Figure 11 is a typical time record as shown on the type 47 film. Time breaks are usually sharp due to the excellent shock-wave transmission in water-saturated sediment.

The Geospace refraction system used yields 12 recording traces and one time break. The recorder provides 100 hertz timing lines (10 milli-seconds) on the record. Recording time of the instrument can be selected as .2, .3, or .4 second. These times allow bedrock deeper than 500 feet to be located, assuming a two-layer case and speeds of 5,000 and 15,000 feet per second for the upper and lower layers, respectively.

Lightweight Geospace geophones with a maximum sensitivity of 14 hertz were attached to the takeout cable at 30-foot intervals. The use of slip-on type connectors required preventing sand grains from being stuck between the connector and the takeout on the wire. Whenever lines were run on dry sand or salt marsh, it was necessary to dig the phones into more com-pact sand or peat. On the low-tide terrace, or beach face of the barrier beaches, and on the intertidal bars and tidal deltas, the phones could simply be pressed into the sand, and with this efficiency as much as 1 mile of beach can be profiled in 1 day using 30-foot phone spacings.

Figure 12 is a diagram for a typical seismic refraction line layout employed in this study. Two 12-phone takeout cables were laid down with a two-phone overlap. By convention, phones are numbered away from the center of the spread. A shot is made at the shotpoint with phones on section A. Then the phones are moved to section B and two more shots are made. This yields a total of four records, two representing sediment returns with a shot near the "12-phone" end and two representing bedrock returns with a shot over 330 feet from the "one" end. When plotted, two times are common to each pair of records; this duplication allows for corrections due to loosening of the sediments surrounding the shotpoint.

Amplifier potentiometers are adjusted on the recorder depending on the gain setting required to eliminate background noise. Background noise includes vehicles, wind, pedestrian traffic, and the surf. A heavy surf can seriously affect coastal seismic operations. This gain setting varied with the proximity of the phones to the shot hole. The GT-2A system has an individual gain control for each channel; phones near the shot can have decreased gain on their amplifiers, and more distant phones can have in-creased gain on their amplifiers.

Problems were encountered in the intertidal area where the insulated parts of the phone takeouts or connectors touched the wet sand. A phenom-enon known as "shot feed" almost obliterated most time traces where several connectors touched the wet sand. This is probably caused by the electric blaster current (90 VDC) being conducted through the salt water in the sand

Figure 9. Geospace GT-2A Portable Refraction Recorder. Records are
made on type 47 Polaroid film in camera at center. Knobs
on upper half of panel control the amplifier pots; OPERATE,
TEST, and SAFETY switches are closest to the operator.
Recorder is wired for cap detonation (lower left) and
connected to the jumper cable (right side).

Figure 10.   Geophone Planted in Soil Below the Sod. Phones are also dug
in under dry sand where a hole 8 to 10 inches deep may be necessary to place the geophone spike in more compact sand. On
water-saturated sediments the phone spike can simply be pressed
into the ground.

16

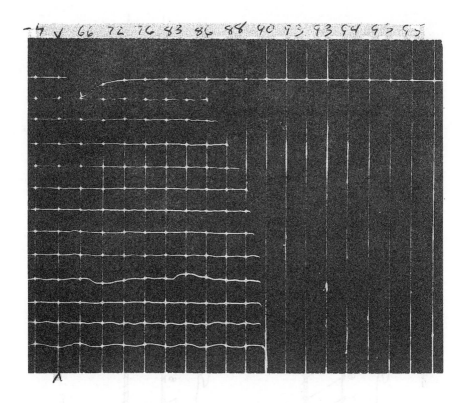

Figure 11. Type 47 Polaroid Film Record. The sharp breaks are typical
of good energy transmission through water-soaked sediments.
The vertical lines are spaced 10 milliseconds apart. The
13 traces include one shot break and 12 geophone channels.
Arrivals can be picked to the nearest millisecond. With
the exception of the "-4" at the left of the caret, the
handwritten numbers are the arrival times in milliseconds.
The "-4" is a constant indicating the shot break's relationship to the initial timing line marked by the caret. Arrivals are timed from the initial time line and the constant
is then substracted from raw times to give real times
written on the record. The almost simultaneous arrival
of the last five phones probably indicates rising bedrock.

17

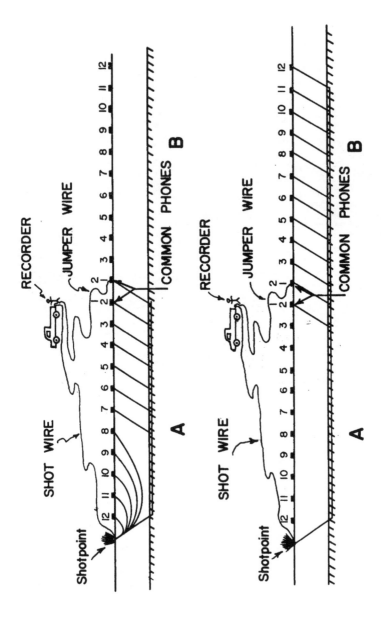

Figure 12. Diagram of Seismic Profile Layout

18

and upsetting the galvanometer whose geophone connector lay in the water-saturated material. This can be avoided by placing connectors on the head of the geophone, off the sand, or by using waterproof connectors to the takeout cable.

Seismic energy was provided by 60 percent gelatin explosive detonated by an instantaneous electric blasting cap. Blasting caps of the seismic variety are recommended due to the larger amount of current required to fire them. Amounts of explosive varied from 1 ounce to 3 pounds; the smaller amounts were used in areas of shallow bedrock and on clam flats; the larger charges were required in dry dune sands. Blasting caps alone or sledge hammer shock sources will not work in coastal sands.

Explosive charges are always buried as deeply as possible to improve energy transmission. This usually means a shovel-dug hole 2 to 3 feet deep on the beach or a bar-punched hole in the marsh peats. The charge should be placed in or near the water table because this allows a reduction in the amount of explosive employed. Dry sand is an energy absorber which suppresses energy propagation unless the explosion can be coupled to the water table.

The ecological effects of explosives in the shore environment require some comment. No more explosive should be used than is required to yield a sharp time break at a particular gain setting, determined by existing background noise. Therefore, three variables are related: background noise, gain setting, and amount of explosive. They should be considered in that order. Naturalists will often express a fear of large scars left by explosion seismology. Such fears are unnecessary because charges up to 1/2 pound can be detonated 3 feet down in marsh peat without disturbing the surface. Seismic charges in dune sand or other dry vegetated areas cause little or no disturbance. Even the larger charges required to pass energy in the loose material rarely leave a scar. Seismic profiles seaward of the high-tide mark on beaches or on intertidal sand bodies without shellfish can be run without undue concern for amount of explosives used. The incoming tide erases the craters.

Shellfish flats do present a problem. Shell fishermen should be informed about the nature of the work. Profiles should be located over areas of sparse shellfish population; gain is then increased and charges as small as 1 ounce can locate a bedrock profile as deep as 40 feet.

Safety is always foremost when using explosives. Only qualified explosive experts should handle and load explosives, or wire the shot holes. The crew should be large enough to keep onlookers from the blasting area, an acute problem in popular beach areas. Blasting caps and dynamite should be stored and transported in separate magazines. If possible, these explosives should be transported in a vehicle used solely for that purpose, rather than in one carrying personnel or seismic equipment, particularly sparking (iron or steel) tools.

III. SEISMIC PROFILE INTERPRETATION

Although time breaks on the type 47 Polaroid film records can be picked
and plotted in the field, it is necessary to replot these data at a larger
scale in the laboratory. Time breaks can be picked only to the nearest
millisecond. A plotting scale of 10 milliseconds per inch on the vertical
axis and 50 feet per inch on the horizontal axis has proved to be a conven-
ient layout for the time-distance plots. After plotting the time points,
a line (or series of lines) is drawn by visual fit through these points.
Velocities for the various layers are computed from these lines. Velocities can be computed to a high degree of accuracy and later rounded.
When drawing these lines through the data points, slopes should be constructed on the time-distance graph in accordance with reasonable seismic
velocities. Topographic relief or subsurface irregularities occasionally
will cause unusually high or low velocities.

1. Formulas and Nomograms

Selection of an appropriate interpretation method for seismic refraction
data has long been discussed by geophysicists. There is a choice between
two formulas: the critical-distance formula, and the time-intercept formula.
The critical-distance formula is an excellent determinant for the classical
two-layer system. However, this formula performs poorly in the multilayer
situation common to the shore environment. The time-intercept formula
(there are several variations) has significant shortcomings, but is more
successful in multilayer situations. For this reason, it is used most
frequently in data interpretation in this study. Nomograms for both
critical-distance and time-intercept methods were obtained. When results
were checked, the calculated thicknesses and depths usually agreed within
5 percent. In addition to the nomograms for depths, a nomogram was also
used to compute offset distances and critical angles. Refraction geophysicists are concerned about the appropriate location of the depth that
has been calculated by the formulas. An offset distance was calculated
as a function of velocity and depth, then the calculated depth was placed
away from the shotpoint by an amount equal to this offset distance.
Figure 13, a sketch of a typical two-layer case, shows the offset distance
with respect to the overlying shotpoint.

2. Seismic Velocities

When determining intercept times or the difference of intercept times
for use in the time-intercept formula, apparent velocities should not be
confused with true velocities. Velocities associated with water-saturated
sediments tend to cluster around 5,000 feet per second with a plus or
minus deviation of about 300 feet per second. Velocities other than those
centered around 5,000 feet per second can be related to the degree of satur-
ation or to the texture of the sediment. Apparent sediment velocities also
may be created in much the same manner as apparent bedrock velocities. This
situation will be discussed later. Apparent and true bedrock velocities

20

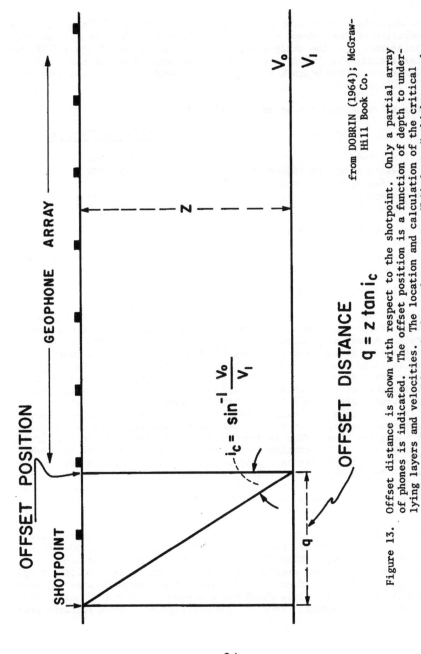

Figure 13. Offset distance is shown with respect to the shotpoint. Only a partial array of phones is indicated. The offset position is a function of depth to underlying layers and velocities. The location and calculation of the critical angle i_c is shown. This angle is used to compute "blind-zone" thickness and offset distance.

from DOBRIN (1964); McGraw-Hill Book Co.

must be separated strictly by statistics and a knowledge of the approximate
configuration of the bedrock surface gained from drill-hole study. Bedrock
velocities vary over the study area, but most are between 15,000 and 19,000
feet per second. The 19,000 feet per second velocities are in the Castle
Neck-Essex area. The bedrock is largely Salem Gabbro Diorite and Cape Ann
Granite (Clapp, 1921).

Seismic velocities can be summarized as follows:

Material	Velocity (feet per second)
dry sand	800-2500
water-saturated sediments	4700-5300
(sand, gravel, and sandy till)	
compact till	6000-9000
bedrock	15,000-19,000

There are situations where a continuous spectrum of velocities exists from
the driest sand to the most compact till. For this reason, it is difficult
if not impossible to ascertain the texture of underlying material by simply
determining its velocity.

3. Graphic Interpretation

    Interpretation techniques and procedures were standarized during the
reduction of the study data. Interpretation of dipping bedrock was studied
in detail. A back projection (on the time-distance plot) from the first
phone showing a bedrock return, using the true velocity of the bedrock,
yielded the time intercept used in the computation of depth. On every
time-distance plot with arrival times from both sediments and bedrock it
is usually obvious which phone received the first bedrock arrival. The
bedrock returns lie along a line of flatter slope, and are usually scattered
about this line in a more random fashion than the sediment arrival times.
In homogeneous sands, it is not unusual for the arrival times to lie pre-
cisely along one line. When in doubt about which phone carried the first
bedrock arrival, it is acceptable to choose the next phone from the shotpoint as the basis for the back projection. Figure 14 shows a back projection of true bedrock velocities on a time-distance plot that might
otherwise be difficult to interpret by the time-intercept method. This
technique can also be used for a dipping till-layer that is yielding an
apparent till velocity rather than a true one.

    Irregular bedrock surfaces are a problem. Relief on bedrock topography
can be computed by a method presented by Meidev (1960) and reproduced in
Figure 15. Empirical results for this study area suggest that use of
1/2 ΔT rather than ΔT may yield better estimates of bedrock relief. Dipping
and irregular bedrock surfaces caused a lack of agreement between adjoining
seismic lines. Seismic lines which share a common shot hole should yield
depths of close agreement. However, where irregular bedrock surfaces pre-

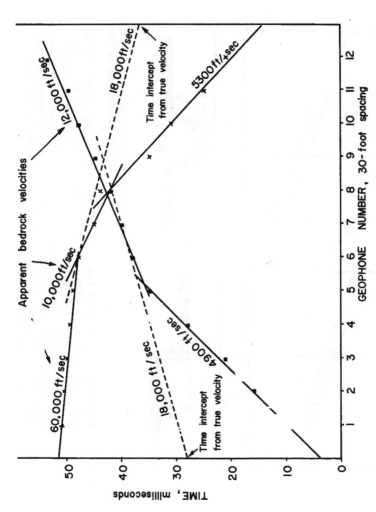

Figure 14. Apparent Bedrock Velocities Cannot be Used to Determine Time Intercepts. True velocities are drawn back to the time axis from the first phone showing a bedrock arrival. The assumption is made that between the shotpoint and the first bedrock phone the bedrock surface is not dipping. This is sometimes an incorrect assumption, but in conjunction with data from adjoining lines yields a good approximation of depth.

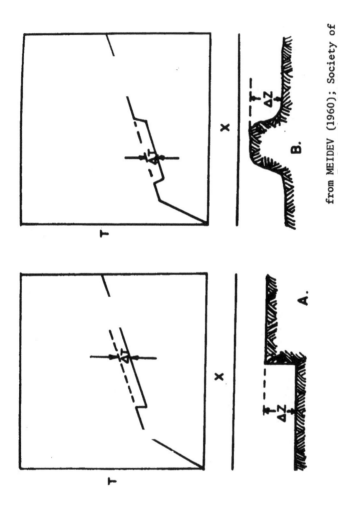

Figure 15.   Meidev (1960) Suggested the Determinations of the Amplitude of Bedrock Topography (ΔZ) by the Use of ΔT From the Time-Distance Plot.
ΔT is treated as a time-intercept quantity and used in the time-
intercept formula.   Sketch shows the time-distance plot over a scarp
(A) and the relationship of a pinnacle to a time-distance plot (B).
Empirical results for this study area indicate that 1/2 ΔT provides better estimates of bedrock relief.

from MEIDEV (1960); Society of
Exploration Geophysicists

dominate, it is not unusual for depths to disagree by 20 to 30 percent
particularly if the lines are oriented at right angles to one another.

4. Problems of Interpretation

The line connecting the travel times for a seismic wave in the first
layer rarely passes through the origin on the time-distance plot. Figure
16 is an example of this non-zero intercept; the figure is a reproduction
of the time-distance plot of seismic line CB-8A east of Crane Beach. The
location of this intercept on this time axis closely matches the depth to
the water table in the sand over the seismic profile. A comparison
between the penetration drill-log for the nearby drill hole CBD and the
time-distance plot showed that the water table is about 3 feet beneath
the surface. The first segment of the time-distance plot intercepts the
time axis at about 6 to 8 milliseconds. Assuming a velocity in dry sand
of about 1000 feet per second for the seismic wave, it can be proven that
this timelag represents a 3- to 4-foot overburden of dry sand.

Previous discussion of the slower velocity of the seismic wave in dry
sand suggested the complications in placing a geophone spread over dunes.
Problems of interpretation can be greatly reduced by seeking a level route
through the dunes for the seismic line. In addition, a sketch of dunes
should be placed on the field notes if the seismic line is not surveyed.
A thickening or thinning of this upper dry layer yields an apparent sediment velocity which can be confusing. Figure 17 is a sketch showing a
3-foot thickening over a 330-foot geophone spread. This thickening can
change the apparent velocity of (5000 feet per second) water-saturated
sediment lying underneath this dry layer. Sediment below the water table
rarely yields a seismic velocity less than that of the range 4700 to 5000
feet per second. If a lower velocity is found, a situation similar to
that in Figure 17 should be sought as an initial explanation. Occasionally
there is a velocity variation beneath the water table which can be attri-
buted only to sediment texture or to the environment of deposition. Such
a horizon could be verified only by drill-hole correlation.

Plum Island seismic lines PI-7, PI-8 and PI-9 and the nearby drill hole
PIC suggest another problem common to seismic work near the shore. Apparent
attenuation of the seismic energy takes place along this profile due to a
possible thickening of a peat sequence beneath a dry overlying sand (Figure
18). The resulting time-distance plot is unusual (Figures 18 and 19).

The presence of peat under a refraction profile does not always create
a time-distance plot with a step-like appearance like that in Figure 18.
Recent work by the author on Cape Cod beaches near Brewster, underlain by
marsh-peat sequences, shows that these layers have little effect on seismic
transmission. In the area of Plum Island and Castle Neck, the presence of
these step-like breaks in the time-distance plot strongly suggests that a
nearby drill hole might show an energy absorbing layer, such as peat, near
the surface.

25

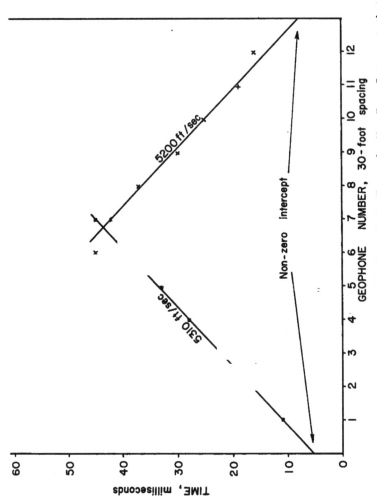

Figure 16. Time-Distance Plot of Seismic Line CB-8A Shows the Non-Zero Intercept that Can be Used to Determine the Depth to the Water Table. Penetration records for a nearby drill hole indicate the water table was 3 to 4 feet below the surface. One-half of the intercept time in milliseconds usually equals the water-table depth in feet.

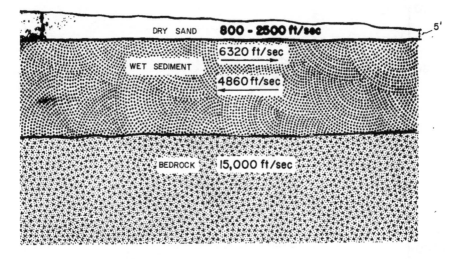

East
Shotpoint

West
Shotpoint

DRY SAND 800 - 2500 ft/sec   5'

6320 ft/sec
WET SEDIMENT
4860 ft/sec

BEDROCK   15,000 ft/sec

Figure 17. Cross Section Determined by Nearby Drill-Hole Data Explains
            Apparent Sediment Velocities. The wet sediment has a true
            velocity of 5,000 feet per second. However, with a dry sand
            overburden of variable thickness, apparent velocities are
            indicated on the seismic record. A high velocity appears
            when the shot is fired at the east end and the seismic wave
            travels upward through a progressively thinner dry sand as
            it propagates westward. However, a shot at the west end of
            the profile, which places energy in the wet sediment, rec-
            ords successively longer times on geophones toward the east
            due to the thickening of the dry overburden. The two appar-
            ent velocities and the respective direction of propagation
            for the seismic energy are indicated on the cross section.

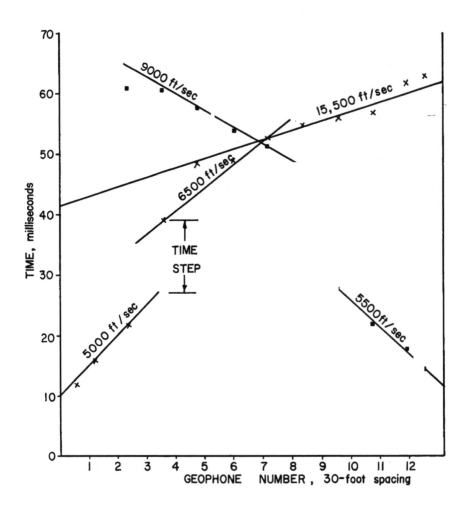

Figure 18. Time-Distance Plot for Plum Island Seismic Line PI-7 Shows
Time Steps that Indicate a Buried Peat Layer

28

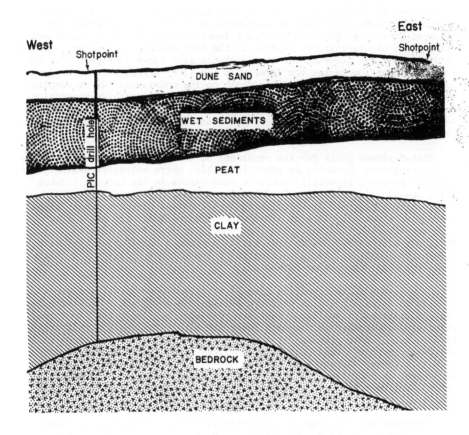

Figure 19. Cross Section from Seismic Data Like that in Figure 18 and Drill-Hole Data from PIC. Energy transmission is slow and poor in the peat. Arrivals below the time step on Figure 18 indicate energy transmitted by sediment above the peat. Arrivals plotted above the time step indicate energy taking a path down through the peat into the clay and returning through the peat from the lower strata. A low seismic velocity of perhaps 1000 feet per second would introduce the time gap indicated by the time-distance plot.

Vincent J. Murphy, of Weston Geophysical Engineers, Inc., (personal communication) mentions similar phenomena in other areas. A condition known as "velocity inversion", where a lower-velocity layer is sandwiched between two higher-velocity layers, is the best description of the causative situation for the steps in the time-distance plot shown in Figure 18. The velocity inversion causes the layer with the lower velocity to be hidden. In addition, an error in computing depth to the underlying layers results. The surface wave reaches a limited number of phones in regular succession, and then attenuates, causing succeeding phones to await the arrival of the wave which must pass through the deeper sands and their overlying peat layer.

Time-distance plots for the remainder of the study area show this step-like arrangement appearing in other localities where extensive peat deposits could be present, especially interior depressions in the barrier islands which now contain brackish or fresh water marshes. Figure 20 shows a time-distance plot and its associated cross section near the Stage Island Pond on southern Plum Island. Peat deposits apparently are associated with the brackish pond and swamp since early stages of the barrier. It is likely that barrier sands now overlie a predune marsh associated with a lower sea level. This is an example of how seismic profiles and nearby drill-hole logs complement each other.

Plum Island seismic profile line PI-17 (transverse profile, Figure 20) is tied to seismic profiles that run parallel to the beach. This arrangement of profiles is desirable wherever possible, because tying three geophone arrays together at a common shotpoint with a right-angle relationship between two of them provides an excellent check on the continuity of seismic returns. In this example, Plum Island lines PI-17, PI-18 and PI-19 are tied to a common shotpoint on the beach. Till was calculated at a depth of 90 feet on profiles run parallel to the ocean, but was found between 75 and 80 feet on a transverse profile. The time-distance plot also shows the characteristic time steps due to the buried peat.

Glaciomarine clay deposits are extensive beneath Plum Island. Their presence was known before this study, in part from a study (McCormick, 1968; McIntire and Morgan, 1963) of previous work in the surrounding areas, and from the occasional outcrop of similar clays in the Ipswich quadrangle (Sammel, 1963). Sjögren and Wager (1969), Swedish geologists, discovered a velocity contrast between thick clay sequences and overlying wet sand sequences during a preliminary engineering study for foundation construction on fjord and river deposits in northern Sweden. Although velocity contrast was expected between glaciomarine clay and wet sediments in this study area, none appeared, even in locations where impressive clay thicknesses underlie estuarine and barrier sands. For example, Figures 21 and 22 show one of the seismic time-distance curves and the cross section from the marsh-capped Middle Ground in the Parker River. Sand deposits greater than 60 feet thick overlie glaciomarine clays that extend to at least 100 feet, the point at which drilling ceased. The time-distance plots show no

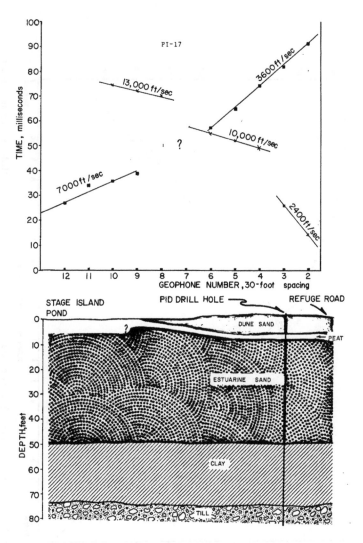

Figure 20. Time-Distance Plot for PI-17 Shows the Time Step Probably
Related to Buried Peat. In addition, till was found in
drill hole PID, and was located on the adjoining seismic
profiles. Till was calculated at a depth of 75 to 80 feet
on the transverse profiles; profiles running parallel to
the beach located till at 90 feet.

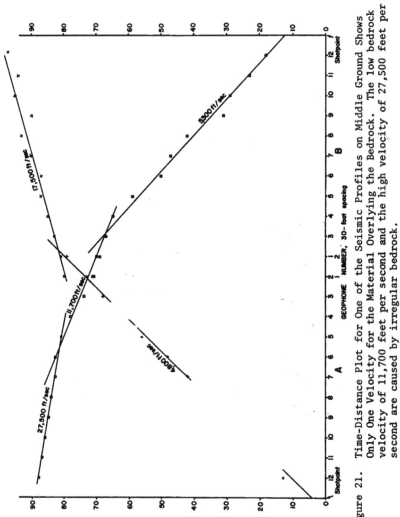

Figure 21. Time-Distance Plot for One of the Seismic Profiles on Middle Ground Shows Only One Velocity for the Material Overlying the Bedrock. The low bedrock velocity of 11,700 feet per second and the high velocity of 27,500 feet per second are caused by irregular bedrock.

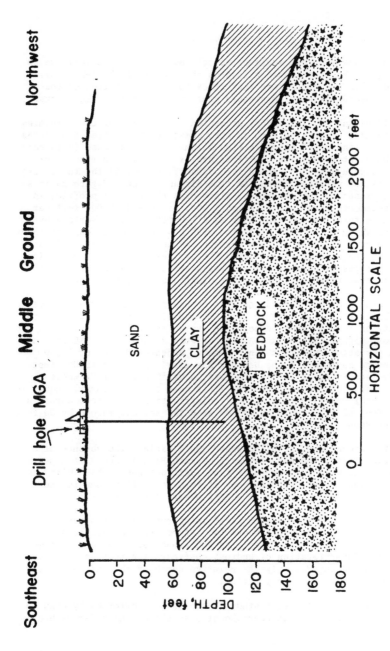

Figure 22. Cross Section of Middle Ground from Both Seismic and Drill-Hole Data Indicates a Slight Bedrock Rise Under the Middle of the Sand Body. Clay is draped over this bedrock rise and estuarine sands compose the upper 60 feet of sediments.

33

velocity contrast for the glaciomarine clay.

The plots show apparent sediment velocities that are probably due to a thickening of the peat sequence towards the center of the sand body. Middle Ground has a bedrock core that rises to within 100 feet of the surface and then drops away to greater depth beneath the present estuary.

Leet (1950) and Hawkins (1960) presented nomographs for the determination of thickness associated with an arrangement of velocities and depths known as the "blind zone" in refraction seismic work. Because it is customary to base interpretation on "first breaks", or first arrivals, of seismic waves at a geophone, there is a subsurface zone in the three-layer case which is not represented on the time-distance plot. Figure 23 is a graphical explanation for the travel times of a three-layer profile, in which the sediments have velocities of 5,000, 8,000 and 15,000 feet per second. These three layers are shown as a bedrock surface overlain by a till layer which is in turn blanketed with a water-soaked sediment layer. Seismic wave rays that take a path through the intermediate layer do not contribute to the time-distance plot. The arrival that took a path along the deeper but faster bedrock surface registers at the geophone before the arrivals that traveled through the intermediate layer.

The thickness of the "blind zone" is dependent on the respective velocities of the three layers (Figure 24). In addition, the amount by which V_3 (velocity of lowest layer) exceeds V_2 (velocity of middle layer) is probably the most significant relationship. Figure 23 shows that 50 feet of 8,000 feet per second sediment can be masked beneath 80 feet of 5,000 feet per second sediment if the bedrock has a velocity near 15,000 feet per second. According to Leet's nomogram, a maximum factor Y, where $Y = H_2$ (thickness of middle layer)/H_1 (thickness of upper layer), equal to .85, can be achieved with this velocity distribution. In other words, as much as 68 feet of till could be masked from the seismic record by 80 feet of overlying sand.

Thus, the presence of a "blind zone" could explain the absence of a discrete velocity for the clay sequence beneath Plum Island. It appears unusual that data were gathered only from locations where this "blind zone" could mask a higher clay velocity, but it is a possibility due to the thick blanket of uniform sand that almost always covers the clays.

Drill-hole control linked with close interpretation can often help to reduce errors. At locations where two interpretations are possible due to a possible "blind zone", only an additional drill hole can answer the question. Interpretation of bedrock configuration remains the same even with a "blind zone". Only the relative depth of the third-velocity surface changes, not its configuration.

A recent seismic reflection study (Duane 1969), performed by CERC to find available sand for beach nourishment also showed the offshore bedrock

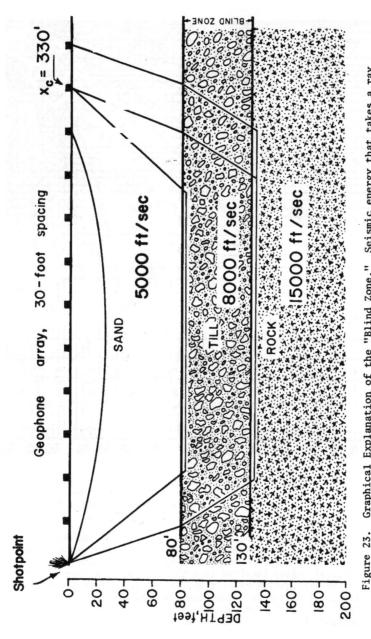

Figure 23. Graphical Explanation of the "Blind Zone." Seismic energy that takes a ray path through the sand and into the till arrives at the same phone at the same time as the energy that passed through both the sand and the till to travel in the faster bedrock. All phones closer to the shotpoint than x_c receive a direct wave through the wet sand; phones beyond x_c receive a wave which included bedrock in its path. Therefore, arrivals from the till layer are masked by those from the bedrock.

35

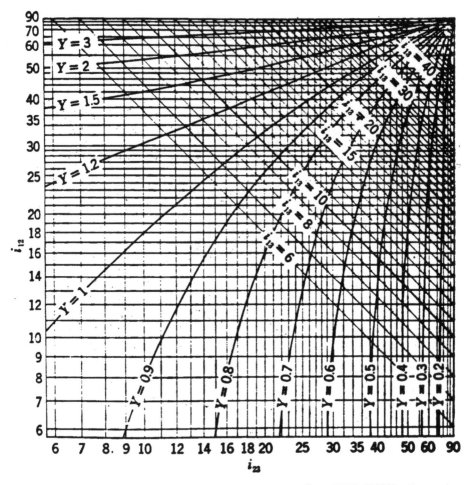

from LEET (1950); Harvard
University Press

Figure 24.   Leet (1950) Presented a Nomogram to Determine the Maximum
Thickness of an Intermediate Layer that Existed as a "Blind
Zone." The nomogram is entered by any two of the three
critical angles involved in a three-layer case (see Figure
13 for explanation of critical angle, i_c). From the nomogram, a Y factor is determined. The thickness of the upper
layer when multiplied by this Y factor yields the maximum
possible thickness for a "blind zone."

topography from Plum Island to Castle Neck. The results were successful, but Duane found that sand bodies might occur in patches because of the unusually rugged bedrock topography, and a more detailed coring program would be necessary to evaluate further the potential sand deposits in the area.

Although insufficient transverse profiles were run to prove the matter conclusively, it appears that a bedrock rise occurs beneath major features such as Plum Island and Castle Neck and smaller forms such as Middle Ground. Figure 21 shows the rise under Middle Ground. Figure 25 summarizes two additional traverses that indicate the possibility of bedrock rises. McIntire and Morgan (1963) found that the northeastern Massachusetts and New Hampshire beaches are anchored on both bedrock and glacial deposits.

The shallower bedrock depths at Crane Beach allowed determination of the shape of that bedrock surface in greater detail than elsewhere. The surface is cut by preglacial depressions and channels into which glacio-marine clay was deposited apparently before the existence of the barrier island. Figure 26 shows a cross section from data of seismic lines CB-18 and CB-19, with the accompanying drill hole CBG. Figure 27 shows the bedrock configuration beneath the Essex flood-tidal delta. This large sedimentary body is apparently sited on bedrock at a shallow depth. Hydrography of the nearby Essex Bay between the rocky headland of Wingaersheek Beach (locally called Coffin Beach) and the southern tip of Castle Neck, indicates that the location of the main channel is restricted by bedrock topography. In addition, the bedrock surface deepens beneath the channel that separates the tidal delta from Castle Neck proper, indicating that this sediment-filled depression once might have contained a deeper Castle Neck River.

Seismic profiles over the swash bars offshore from the Parker River inlet indicate a deeper bedrock horizon. These profiles are without drill-hole control, but even with the possibility of a "blind-zone" error there is unquestionably a great deal of recent sediment without a bedrock or glacial rise on which to anchor these sediments. Profiles such as these on intertidal sand bars in the nearshore zone suggest that seismic refraction methods could aid in the data collecting for a sand inventory program. In addition, sedimentary horizons detected by seismic methods in deeper water could be correlated with similar horizons beneath inter-tidal and exposed shore formations.

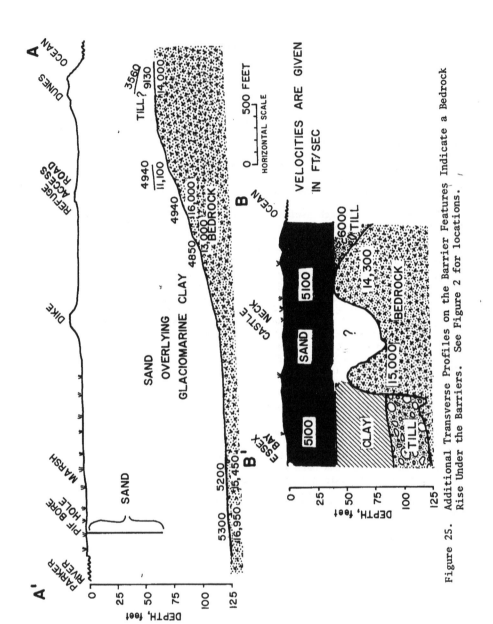

Figure 25. Additional Transverse Profiles on the Barrier Features Indicate a Bedrock Rise Under the Barriers. See Figure 2 for locations.

38

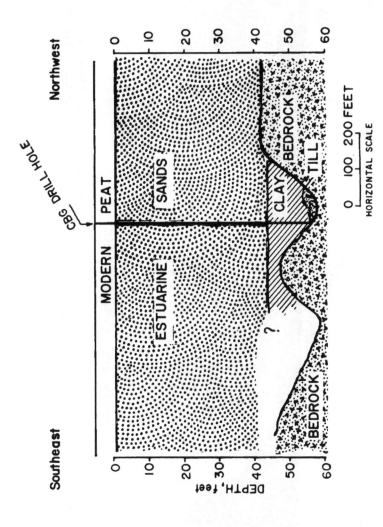

Figure 26. Cross Section from CB-18, CB-19 and Drill Hole CBG

39

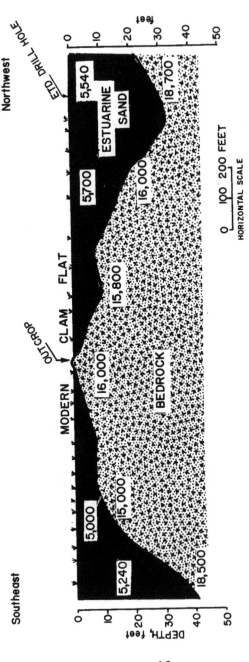

VELOCITIES ARE GIVEN IN FT/SEC

Figure 27. Profile Across the Essex Flood-Tidal Delta Shows that this Sand Body is
Anchored on a Bedrock High. The profile is at drill hole ETD on Figure 2.

40

1. Generalized Stratigraphy

Sampling techniques have been discussed in Section II. Samples were given a preliminary inspection in the field upon removal from the split-spoon sampler. This allows the driller to make decisions about sampling interval and perhaps estimate the proximity of a significant sedimentary change.

The Drill-Log Appendix summarizes the preliminary field observations and later binocular microscope analyses of the samples. Using the binocular microscope, estimates were made of grain size, sorting ,roundness, and gross mineralogy.

Examination showed a distribution of sediment that closely agrees with previous work of McIntire and Morgan (1963) and McCormick (1968). This study went beyond the depths of the McCormick study and added to the interpretation of the geological history of Plum Island. The paleogeography suggested by this study fits well with the ideas about submerged relict dunes expressed by Anan (1971).

McCormick determined seven stratigraphic zones by analysis of marsh cores. His generalized column follows:

High salt-marsh peat
Spartina alterniflora peat
Western fine-grained facies
Eastern coarse-grained facies
Black peat
Weathered zone
Blue clay

McCormick's samples were studied by x-ray and grain-size analyses. Such analyses were not made in this study, because it is more concerned with the large-scale, more obvious, correlations. For example, results of this study allow broader correlation (extension deeper and further seaward) of such horizons as the weathered zone, which overlies the glaciomarine clay sequence in McCormick's column.

2. Glacial Till and Glaciomarine Clay

Overlying the igneous bedrock surface is glacial till, which varies from highly compact drumlin material to extremely sandy till beneath some of the beach deposits. Figure 28 is a photo of this sandy till from hole CBB. Identification of this material as till is sometimes disputable; the possibility that it is outwash or an ice-contact deposit is acknowledged. However, the material is poorly sorted, angular, and the pebbles have a silt cap-evidence that indicates till. The

Figure 28. Sandy Till from Drill Hole CBB is Typical of the Coarse Glacial Material Brought up in the samplers. A silt cap is always present on the pebbles. A silt cap is a thin, silt, clay layer adhering to the upper surface of the clasts.

42

compact till is identical to that of the major drumlins in the area. Coarse material is tightly packed with a matrix of fine silt-clay-sized particles. In fact, without the coarse angular pebbles, the fine part of this till could be confused with the glaciomarine clay when viewed in a small quantity in a sampler.

The absence of glaciomarine clay overlying the sandy till in some drill holes (CBC, CBD and CBE) suggests that the clay was never deposited here, or that if deposited, it was eroded during a later transgression by the sea. I believe that the clay was deposited on top of glacial material as found beneath Crane Beach, but was removed by erosion. Perhaps the sandy till, which is so common in the Crane Beach drill holes, is simply the compact till reworked by the sea and mixed with coastal deposits. If a deeper drilling had been possible, the sandy till might have turned out to be a veneer over the more compact till.

The glaciomarine clays were deposited during a high stand of the sea in a blanket of variable thickness over existing topography, lapping up on the sides of drumlins. The presence of glaciomarine deposits in a steep bluff at the southern end of Plum Island above sea level, and at elevations higher than 40 feet near Ipswich, contribute to the uncertainty about how these deposits were superimposed on the topography. Glaciomarine clay usually appears at depths below 50 feet (see Table); the same clay outcrops are throughout the Ipswich quadrangle (Sammel, 1963). This gap between surface outcrops and the 50-foot depth may be an erosional situation. More observations of buried clay horizons are needed to determine the configuration of the top of this deposit.

3. Indurated Clay

An indurated layer is found on top of the clay layer (see Table). Five of 10 bore holes containing marine clay showed an indurated layer. This layer is a hard cement-like material that is 3 to 10 inches thick. Once the material has been loosened by the chopping bit it closely resembles the softer clays below. The layer is located by the difficulty of drilling through it. The hard clay is washed up the holes in chips or lumps, and becomes softer as it is mixed with water.

Such a layer may be dewatered by compaction. This is a common phenomenon, especially in areas of artificial fill overlying clay deposits. Induration may also be due to exposure to subaerial weathering. An alternate mechanism, secondary precipitation, has been suggested by Coleman and Ho (1967).

Coleman and Ho studied recent sedimentary sequences in the Atchafalaya Basin, Louisiana. They obtained undisturbed cores of 120-foot borings near the Mississippi Delta. The age of the deposits ranged from 10,000 B.P. to present. This extensive study determined that increase in compressive strength and decrease in water content in sediments in this area was partly due to the presence of cementing minerals such as iron hydrous

43

TABLE-BORE HOLES CONTAINING MARINE CLAY

Hole	Indurated Layer	Sandy Till Encountered	Depth in Feet		
			To Clay	To Clam Flat	Total of Hole
PIA	Yes	---	60	--	100
PIB	Yes	---	61	--	100
PIC	Yes	---	21	15	60
PID	Yes	Yes	50	15	77
PIE	Yes	---	45	34	100
PIG	?	---	66	22	100
MGA	No	---	60	46	100
CBB	No	Yes	58	--	59
CBF	No	Yes	72	54	84
CBG	No	Yes	48	--	63

oxides, calcium carbonates, minor amounts of siderite, manganese oxides, and manganese carbonate, with additional strength contributed by accumulation of fine clay fractions after burial.

A similar process may have affected the Plum Island sediments. The possibility presents a question that could be answered by further study. An undisturbed core of the indurated layer is needed. Although the cementing agents mentioned may be present deeper in the glaciomarine clay sequence, it is possible that there is a zone of greater accumulation at the top of the clay. Unquestionably, carbonates exist in the blue clay sequence, because calcite nodules can be seen and a positive carbonate reaction to hydrochloric acid is common.

4. Weathered Zone

The weathered zone above the glaciomarine clay is believed to rest on a surface of deposition on which clastic materials were deposited during a low stand of the sea. This weathered zone is composed of coarse, poorly sorted and highly oxidized sand and gravel. Figure 29 is a photo of this material from Plum Island hole PIG. These clastics could come from nearby drumlin tills, ice-contact drift, or stranded beach deposits of earlier sea levels.

5. Black Peat

The black peat here is likely a correlative of the black peat found by McCormick in his drill holes further north. Johnson (1925), Davis (1910) and Bloom (1968) all recognized this peat in New England marshes. These investigators agree that the black peat represents an accumulation of fresh- and salt-water plants deposited in the zone of transition from salt marsh to upland vegetation. The peat is a highly organic compacted layer. Its dark color is probably due to a reducing environment established after transgression of the sea. Its silt content is variable. It ranges in thickness from 1 foot to a few inches, and is totally absent in some of the more seaward drill holes. The black peat does not overlie clay everywhere, because this transitional situation probably could not develop in lower or more seaward locations. This peat layer commonly lies directly under a thick estuarine sequence of either channel, point-bar, or low energy mud-flat deposits rather than a marsh deposit. It is found directly on the weathered zone when all of these layers are present.

6. Estuarine Sediments

The sands that overlie the glaciomarine clay sequence and its capping weathered zone represent various environments of deposition. Most of the subdivisions grouped under the heading of estuarine sediments that are only drill hole horizons, not seismic horizons. Occasionally the seismic record shows a horizon within these sediments, but detailed stratigraphy can only be accomplished with closely spaced drill holes.

45

Figure 29. Weathered Zone is Composed of a Sand and Gravel Mixture that Has an Iron-Stained Color. The pebbles are subrounded to angular. The material is poorly sorted.

46

Clam-flat sequences overlying the clays are often thick (see log for drill hole PID in the Appendix). The clam-flat facies are identified by a large accumulation of clam shells and by an extremely silty to muddy nature.

Figure 30 is a sketch of the study area near Castle Neck as it might have looked before the readvance of the sea over the glaciomarine deposits. Stranded beach ridges probably contributed the material to the weathered zone that appears over the glaciomarine clay. Glacial material provided topographic highs on which to anchor the modern barrier sands.

It is suggested that the Essex Bay flowed in nearly the same channels as it does at present. This suggestion is supported by the fact that the Essex flood-tidal delta is anchored on bedrock. The bedrock topography is glacial or preglacial, and this tidal delta has apparently been stable throughout the sea-level rise. The existence of a deep bedrock low between the Essex delta and the modern Castle Neck suggests that a deeper relict Castle Neck River might have flowed in this preglacial-bedrock low, and built the thick sequence of sediments presently filling it. Glaciomarine clay is found in the bedrock low at the margins of the modern Castle Neck. Its depth varies between 50 feet (log of CBF) and 70 feet (log of CBG). An extensive clam-flat sequence overlies the clay horizon in drill hole CBF, which suggests a fringing clam-flat environment behind a topography of glacial material during the low stand of sea level.

Other drill holes support the scheme in Figure 30. For example, drill hole CBB bottomed in till at 60 feet. Directly on top of the till is a brown clay that grades upward into a blue-green clay before becoming a brown clay again. It is presumed that some weathering changes have been preserved in this record. A highly organic, perhaps fresh-water, black peat overlies the clays. Coarse sand to fine gravel predominates in the section upward from this point.

Drill hole CBB is on the Ipswich River west of Castle Hill. It is one of the more inland sites, and its record suggests that much geologic history could be discovered by a comprehensive deep drilling program in some of the more protected or inland marshes. Drill holes at such areas could sample the black-peat horizons for age dating purposes and determine more closely the sea-level curve for this coastline.

V. CASE HISTORIES FROM PLUM ISLAND

Extensive sand deposits underlie the Parker River estuary. For example, the log of wash-boring MGA (Middle Ground) shows more than 55 feet of sand on top of the clay horizon. Only the lower 10 to 15 feet of sand is thought to be clam-flat deposits. The upper part of the deposit, fine quartzose sand interspersed with organic layers, is thought to have been deposited on sand flats dominated by strong currents and rapid sediment transport. Shellfish do not thrive in such environments on the modern

47

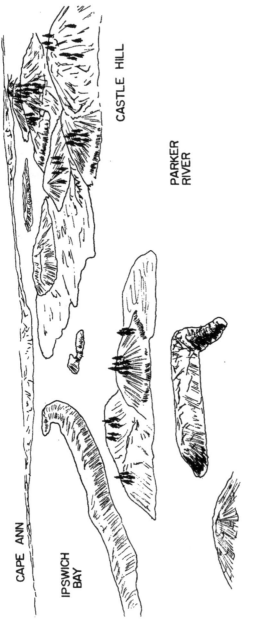

CAPE ANN

IPSWICH BAY

CASTLE HILL

PARKER RIVER

Figure 30. Sketch Looking Southeastward Over the Castle Neck Area. This shows how it might have looked during lowered sea level (10,000 B.P.). Glacial material provided topographic highs against which the modern barrier became anchored.

48

Middle Ground (Boothroyd, personal communication, 1971). Therefore, Middle Ground had its beginnings as a clam flat on a surface of glaciomarine clay, which is anchored on a slight bedrock rise (see Problems of Interpretation, Section III, 4), and changed to a rapidly accreting mid-channel bar as the estuary continued to fill.

Drill hole PIF shows a thick sequence of estuarine sands from -65 feet up to a buried peat layer at -28 feet and then general coarsening from there upward. The stratigraphy determined by McCormick (1968) agrees with this sequence. McCormick found that the western fine-grained facies (a fine, mud-flat sediment) has been replaced and overlain by the coarser eastern coarse-grained facies (a coarse channel sediment) as the estuary migrated landward in response to barrier transgression and sea-level rise. Between about -38 feet to -48 feet there are alternating sandy and muddy layers in drill hole PIF. The coarser, more angular, more feldspathic sediment above the peat layer perhaps represents a higher energy environment, that is, the encroachment of the major channel of the estuary. The drill hole is located on the eastern margin of the estuarine channel where the marsh peat is eroding in response to the meandering of the Parker River channel, which apparently moved westward over this spot centuries ago. Cottages on the channel are rapidly losing their foundations. Historical accounts by previous inhabitants affirm the modern eastward movement of the estuarine channel.

Drill hole PIF went to a depth of 65 feet. No clay was found, but nearby drill holes indicate that clay should not be far below the 65 feet horizon. Deeper bedrock topography probably accounts for the lower clay horizon. Insufficient drilling depth left a small, but perhaps significant, question unanswered.

Drill logs for holes PIA and PIB are nearly identical. In both holes, grain size varies upward from the clay horizon. PIB shows a coarse gravel or weathered layer above the indurated clay; the weathered material probably became the floor of the developing estuary. The alternating silt and sand layers noted in PIF are also present in PIB, confirming that the main channel of the estuary migrated back and forth significantly through time. The peat layer at -11.5 feet in PIB is probably salt-marsh peat that has been compressed by advancing barrier dune sand.

Figure 31 is a sketch drawn from seismic data and drill-hole logs of holes PIC, PIE and PIG. These drill holes were selected for correlation due to their proximity and interrelated depositional sequence. It is suggested that during the low stand of the sea, after deposition of the glaciomarine clays, the estuary maintained a major channel east of its modern change. The original barrier then was farther seaward (Anan, 1971), and the estuary surrounded the drumlins at the southern end of Plum Island. McCormick's map of sediment distribution agrees with a more eastward location of the younger estuary. Clams flourished in the low-energy fringing mudflats along the irregular drumlin shoreline. While

49

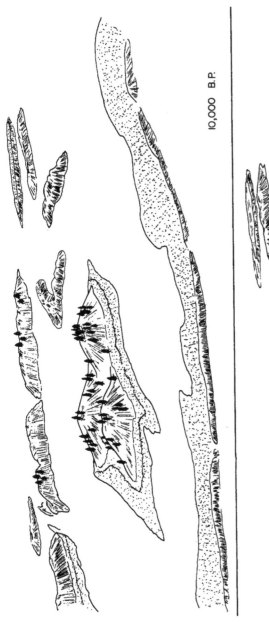

10,000 B.P.

4,000 B.P.

Figure 31. Sketch of Southern Plum Island Looking West During Lowered Sea Level (10,000 B.P.) Shows Drumlins in an Estuary Behind a Low Barrier. The lower part shows sea level at about 4,000 B.P. The drumlins are eroding as a marsh is developing in the small embayment behind them. This embayment will later become a brackish pond.

50

clam flats existed in the vicinity of the drumlins, channel sands were being deposited elsewhere. As sea level rose and the barrier island transgressed, the glaciomarine and glacial deposits were covered with barrier and estuarine sands. However, the accretion was slow enough to allow continuation of the shellfish population on the clam flats.

As the waves began to attack the drumlins just seaward of the relict estuary, a sand spit built across the mouth of the channel behind the drumlins and closed off this estuarine circulation. With estuarine circulation cut off from the basin, marsh peat began to develop. The fresh-water peat, identified in drill-hole log PIC, occurred during a brief fluctuation in sea level or during the complete closure of the basin away from the estuary. The result is a modern brackish pond which is now maintained by a manmade dike.

The logs of PIC and PIG show unusually high elevations for blue clay in a Plum Island drill hole. PIC was drilled 60 feet to refusal, which was determined by seismic profile to be the surface of bedrock. This is an example of the marine clay being draped over a preglacial bedrock high with little later erosion. However, the log of PIG shows silt at 22 feet. On top of the uppermost silt horizon is a gravelly layer with angular rock fragments, typical of the weathered zone over the glaciomarine clay. Glaciomarine clay does not appear until 65 feet. Between these depths is a sequence of alternating clay-silt and sand layers with a semi-indurated layer at 32 feet. The sand and silt are highly quartzose. The blue clay, where it grades into silt-sized material, is usually also highly quartzose. Therefore a sharp discrimination between the silt and the clay is often difficult.

Drill hole PIH is the most southerly hole on Plum Island. At 25 feet, the drill bit passed into poorly sorted, coarse, angular gravel with a silt cap on the pebbles. However, seismic returns indicate a discontinuous "till" velocity beneath this area. If there is a till sequence here, it is likely a thick one on which the southern end of Plum Island is anchored. In contrast, the presence of the mouth of a large inlet also suggests that there could be a large accumulation of coarse channel deposits. A compromise between the two ideas might be that the surface till of southern Plum Island extends southward beneath a veneer of sand deposited by spit accretion. This glacial material has been greatly reworked by wave and current action, but remains near its source, the southern Plum Island drumlin field.

Fitting undated drill-hole logs to sea-level curves can be arbitrary. However, the presence here of such environmental determinants as clam-flat facies, fresh-water peat, weathered gravel zones, and glaciomarine clays permits a more certain three-dimensional space correlation with time. To aid this correlation, the sea-level curves proposed by McIntire and Morgan (1963) and Kaye and Barghoorn (1964), are presented in Figures 32 and 33. Kaye and Barghoorn assume a high stand of the sea before 14,000 B.P.; McIntire and Morgan suggest that the high stand was before

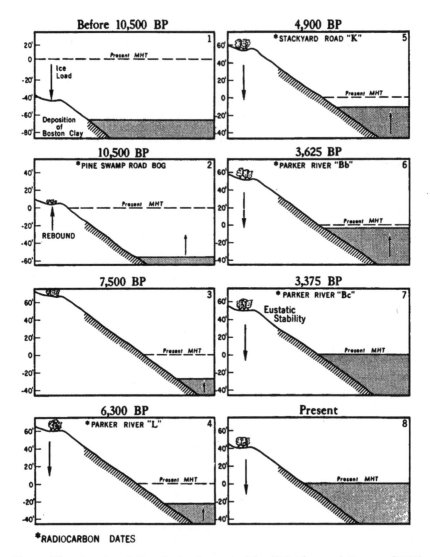

Figure 32. Sea-level Trends as Presented by McIntire and Morgan (1963)
for the Plum Island Area. The levels shown here are based
on radiocarbon dates. Matching the stratigraphy determined
in the above study with that presented by McCormick (1968)
suggests an approximate time frame as in Figures 30 and 31.

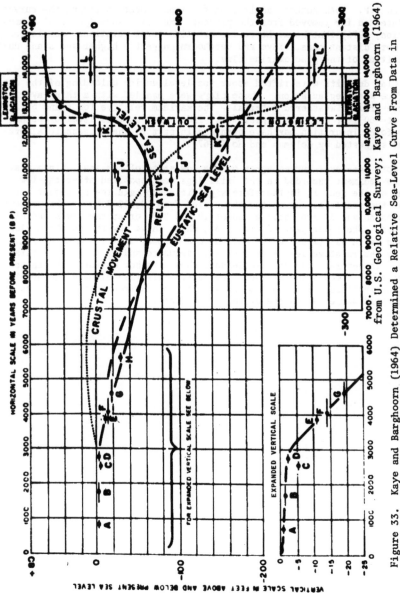

Figure 33. Kaye and Barghoorn (1964) Determined a Relative Sea-Level Curve From Data in the Boston Area. This curve agrees well with that of McIntire and Morgan (1963) for the Plum Island area.

53

10,500 B.P. Both authors admit room for improvement in fixing the curves to dates so far removed from the present. McIntire says (personal communication, 1971) that he might revise his time-level relations, particularly the more distant ones. Kaye and Barghoorn suggest a higher sea stand during deposition of the glaciomarine clay.

Clay deposits were lifted above sea level during unloading of the earth's surface with deglaciation. These clays were mantled with a thin layer of coarse-grained weathered material. The maximum relative low stand of the sea at -60 feet does not explain clay induration at this level or below. Therefore, induration may have been due in part to the diagenetic alterations suggested by Coleman and Ho (1967).

Both sea-level curves agree on the low stand of the sea. Since most of the sediment recorded in the drill logs was deposited after the low stand of the sea, it is therefore significant to make the agreement between these two sea-level curves for the period from 10,000 B.P. to the present. Figures 30 and 31 are dated as about 10,000 B.P. or at the low stand of the sea. The lower part of Figure 31, which shows how the southern Plum Island basin was closed off, could have a date of about 4,000 B. P., due to the relative levels of the marsh sequence and of the sea at this time. In general, the drill-hole data show a rather uniform rise in sea level after a low stand at about -60 feet. This rise allowed the development of shellfish flats in low-energy fringing environments as well as the more rapid accretion on the dynamic sand flats.

VI. DEVELOPMENT OF A BARRIER ISLAND

The findings of this study give all major barrier-island theories some support. As early as Dana (1894) and Gilbert (1890) littoral transport was considered a valid barrier-building mechanism. Johnson (1925) suggested that wave-cut drumlins provided sediment for the development of spits and pocket beaches in the Nantasket area near Boston, Massachusetts. Figure 34 is a reproduction of a similar suggestion by Nichols (1941) for the development of Castle Neck. Nichols did not have the benefit of subsurface data, and was unaware of the Pleistocene topography of the southeastern end of Castle Neck.

Hoyt (1967) proposed relict beach ridges as the origin of barrier islands. Fisher (1968) suggested a mechanism similar to that of Gilbert's (1890) spit-accretion. There are in this study area drowned relict drumlins linked by deposits caused by littoral drift. However, the erosion of the drumlins was not the only source, because the Merrimack River was contributing large amounts of sediment during the building of the barrier islands and spits.

The remainder of Plum Island (north of Camp Sea Haven, Figures 1 and 2) which was not part of this study area, may have a structure similar to the southern end of the island. It is suggested that Pleistocene topography and bedrock highs anchor the island at these locations. Anan (1971) finds evidence in his offshore sediment analysis of a relict delta near the mouth of the modern Merrimack River. The delta was flanked by low barriers which probably moved landward in response to the sea's transgression. A three-dimensional analysis of other barriers on this coast would certainly expand the historical and stratigraphic concepts of the Pleistocene and Holocene sediments.

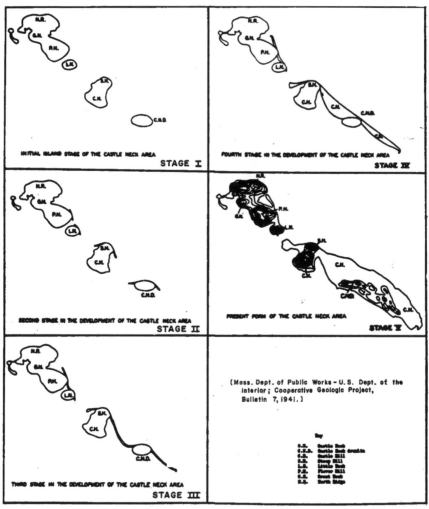

Figure 34. Nichols (1941) Suggested that Castle Neck Originated
From a Series of Beaches and Spits Building Around
Drumlins. His ideas seem quite accurate when com-
pared with subsurface data.

56

VII. CONCLUSIONS

Concluding remarks deal with two separate phases of the study. The first six items summarized below deal with the development of field techniques; the remaining items deal with geologic interpretations resulting from this study.

1. A seismic refraction study should precede the drill-hole study if both are to be done.

2. Drilling equipment that is capable of reaching the deepest bedrock regardless of the nature of the overburden should be selected.

3. A high explosive should be used as the seismic energy source to avoid signal attenuation.

4. Preliminary time-distance plotting should be done in the field using a nomogram solution to gain an accurate judgement about the location of future profile lines.

5. Dunes or large accumulations of dry sand should be avoided when running seismic profiles.

6. Shallow-refraction seismic methods are successful near the shore if attention is given to the following problem areas:

   a. The "blind zone" due to a second layer of intermediate velocity.

   b. The non-zero intercept due to a dry layer between the surface and the water table.

   c. Apparent sediment velocities due to thickening or thinning of a low-velocity layer.

   d. Attenuations of seismic energy in a layer underlain by peat.

   e. Apparent bedrock velocity due to either a highly irregular surface or a steeply dipping surface.

7. Seismic and drill-hole results suggest a well-defined bedrock surface. Bedrock topography is highly irregular and varies from surface outcrop to depths in excess of 150 feet. Mantled on this bedrock surface is till and many Pleistocene topographic highs. This Pleistocene topography can be inferred largely from the modern drumlin configurations, although some glacial topography has been reworked by the sea or buried by coastal deposits.

8. Deposited on this glacial drift is a thick layer of glaciomarine clay. The clay, of nonuniform thickness, appears to cover most

57

glacial topography that was submerged during the time of clay deposition. The clay is not encountered at one particular elevation, but rather resembles a cloth draped over an uneven surface. Clay horizons in the local area vary from 40 feet above sea level to 60 feet below sea level.

9. During the subsequent low stand of the sea, this glaciomarine clay was exposed to subaerial conditions and a weathered horizon consisting of weathered clay and coarse clastic material eroded from nearby Pleistocene topography. In addition, fresh-water vegetation in areas beyond tidal waters developed a fresh-water peat. An approximate time frame for this low stand of the sea is about 10,000 B.P. (Kaye and Barghoorn, 1964).

10. As sea level rose relative to the land, estuarine and marsh sediments accumulated behind a landward migrating barrier. Drumlins were eroded, some so much that they are barely distinguishable today. Finally the sea-level rise leveled off to its current rate of 0.3 feet per century (McIntire and Morgan, 1963), and the barriers became anchored against both buried and exposed Pleistocene deposits.

LITERATURE CITED

ANAN, F.S., "Provenance and Statistical Parameters of Sediments of the Merrimack Embayment, Gulf of Maine," Unpublished Ph.D. Thesis, University of Massachusetts, Boston, Mass., 1971.

BLOOM, A.L., "Postglacial Stratigraphy and Morphology of Central Connecticut," *Guidebook for Field Trips in Connecticut*, New England Intercollegiate Conference, New Haven, Conn., 1968, 305 pp.

BOOTHROYD, J.C., Personal Communication, 1971.

CHUTE, N.E., and NICHOLS, R.L., "Geology of Northeastern Massachusetts," Bulletin No. 7, Cooperative Geology Project, Massachusetts Department of Public Works and U.S. Geological Survey, 1941.

CLAPP, C.H., "Geology of the Igneous Rocks of Essex County, Massachusetts," Bulletin 704, U.S. Geological Survey, Washington, D.C., 1921.

COASTAL RESEARCH GROUP, *Coastal Environments*, N.E. Massachusetts and New Hampshire Field Trip Guidebook for the Society of Economic Paleontologists and Mineralogists, University of Massachusetts, Boston, Mass., 1969, 462 pp.

COLEMAN, J.M., and HO, C., "Early Diagenesis and Compaction in Clays," *Proceedings of First Sumposium on Abnormal Subsurface Pressure*, School of Geology and Department of Petroleum Engineering, Louisiana State University, 1967, pp. 23-50.

CURRIER, L.W., "The Seismic Method in Subsurface Exploration of Highway and Foundation Sites in Massachusetts," Circular 426, U.S. Geological Survey, Washington, D.C., 1960.

DANA, J.D., *Manual of Geology*, American Book Co., 1894, 1,087 pp.

DAVIS, C.A., "Salt Marsh Formation Near Boston and Its Geological Significance," *Economic Geology*, Vol. 5, 1910, pp. 623-639.

DOBRIN, M.B., *Introduction to Geophysical Prospecting*, 2nd ed., McGraw-Hill, New York, 1960, 446 pp.

DUANE, D.B., "A Study of New Jersey and Northern New England Coastal Waters," *Shore and Beach*, Oct. 1969.

FISHER, J.J., "Barrier Island Formation Discussion, *Geological Society of America Bulletin*, Vol. 79, 1968, pp. 1,421-1,426.

FOLK, R.L., *Petrology of Sedimentary Rocks*, 2nd ed., University of Texas, 1968, 190 pp.

GILBERT, G.K., "Lake Bonneville," Monograph 1, U.S. Geological Survey, Washington, D.C., 1890.

59

HAWKINS, L.V., and MAGGS, D.,"Nomograms for Determining Maximum Errors and Limiting Conditions in a Seismic Refraction Survey With a Blind-Zone Problem, *Geophysical Prospecting,* Vol. 9, 1966, pp. 526-532.

HOYT, J.H., "Barrier Island Formation," *Geological Society of America Bulletin,* Vol. 78, 1967, pp. 1,125-1,136.

HOYT, J.H., "Barrier Island Formation: Reply," *Geological Society of America Bulletin,* Vol. 79, 1968, pp. 1,427-1,432.

JOHNSON, D.W., *Shore Processes and Shoreline Development,* Wiley, New York, 1919, 584 pp.

JOHNSON, D.W., *New England-Acadian Shoreline,* Wiley, New York, 1925, 608 pp.

KAYE, C.A., and BARGHOORN, E.S., "Late Quaternary Sea-Level Change and Crustal Rise at Boston, Massachusetts, With Notes on the Autocompaction of Peat," *Geological Society of America Bulletin,* Vol. 75, 1964, pp. 63-80.

LEET, L.D., *Earth Waves,* Monographs in Applied Science, No. 2, Harvard University Press, 1950, 122 pp.

LINEHAN, D., "Seismic Prospecting in New England," *Transactions, 23rd Annual Meeting of the American Geophysical Union,* Pt. 2, 1942, pp. 227-228.

LINEHAN, D., "Seismology as a Geologic Technique," *Application of Geology and Seismology to Highway Location and Design in Massachusetts,* Highway Research Board Bulletin 13, National Research Council, 1948, pp. 77-85.

McCORMICK, C.L., "Holocene Stratigraphy of the Marshes at Plum Island, Massachusetts," Unpublished Ph.D. Thesis, University of Massachusetts, Boston, Mass., 1968.

McINTIRE, W.G., and MORGAN, J.P., "Recent Geomorphic History of Plum Island, Massachusetts, and Adjacent Coasts," Coastal Studies Series No. 8, Louisana State University, New Orleans, La., 1963.

MEIDEV, T., "Nomograms to Speed Up Seismic Refraction Computations," *Geophysics,* Vol. 25, 1960, pp. 1,035-1,053.

POWERS, M.C., "A New Roundness Scale for Sedimentary Particles," *Journal of Sedimentary Petrology,* Vol. 23, 1953, pp. 117-119.

SAMMEL, E.A., "Surficial Geology of the Ipswich Quadrangle, Massachusetts," Geological Quadrangle Map GQ-189, U.S. Geological Survey, 1963.

SEARS, J.H., "The Physical Geography, Geology, Mineralogy, and Paleon-
tology of Essex County, Massachusetts," Essex Institute, Salem, Mass.,
1905.

SJÖGREN, B., and WAGER, O., "On a Soil and Ground Water Investigation With
the Shallow Refraction Method," *Engineering Geology,* Vol. 3, 1969, p. 61.

APPENDIX

DRILL-LOG SUMMARIES

    The logs of 17 drill holes follow. The information is only a summary of each log.

    Preliminary field data and the detailed descriptions from a binocular microscope in the laboratory- are included at the right of the column; the tick mark indicates the center location of the sample taken.

    Grain size was estimated visually by comparison with sieved sediments of known size. Sorting was estimated from the sorting chart of Folk (1968) and the roundness chart of Powers (1953). Depth in feet is logged to the left of the column.

63

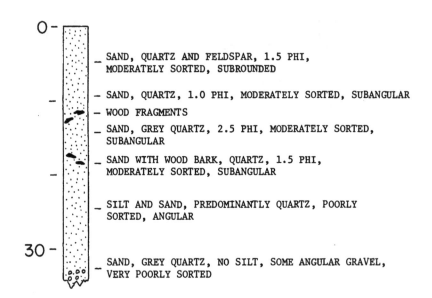

0 —

— SAND, QUARTZ AND FELDSPAR, 1.5 PHI,
MODERATELY SORTED, SUBROUNDED

— SAND, QUARTZ, 1.0 PHI, MODERATELY SORTED, SUBANGULAR

— WOOD FRAGMENTS

— SAND, GREY QUARTZ, 2.5 PHI, MODERATELY SORTED,
SUBANGULAR

— SAND WITH WOOD BARK, QUARTZ, 1.5 PHI,
MODERATELY SORTED, SUBANGULAR

— SILT AND SAND, PREDOMINANTLY QUARTZ, POORLY
SORTED, ANGULAR

30 —

— SAND, GREY QUARTZ, NO SILT, SOME ANGULAR GRAVEL,
VERY POORLY SORTED

CBE

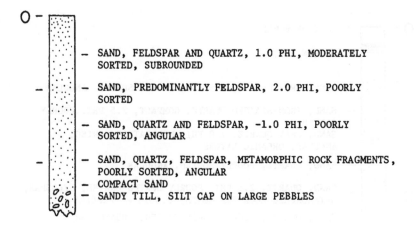

SAND, FELDSPAR AND QUARTZ, 1.0 PHI, MODERATELY
SORTED, SUBROUNDED

SAND, PREDOMINANTLY FELDSPAR, 2.0 PHI, POORLY
SORTED

SAND, QUARTZ AND FELDSPAR, -1.0 PHI, POORLY
SORTED, ANGULAR

SAND, QUARTZ, FELDSPAR, METAMORPHIC ROCK FRAGMENTS,
POORLY SORTED, ANGULAR
COMPACT SAND
SANDY TILL, SILT CAP ON LARGE PEBBLES

PIH

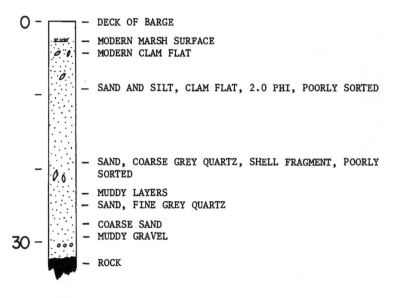

DECK OF BARGE
MODERN MARSH SURFACE
MODERN CLAM FLAT

SAND AND SILT, CLAM FLAT, 2.0 PHI, POORLY SORTED

SAND, COARSE GREY QUARTZ, SHELL FRAGMENT, POORLY
SORTED
MUDDY LAYERS
SAND, FINE GREY QUARTZ

COARSE SAND
MUDDY GRAVEL

ROCK

ETD

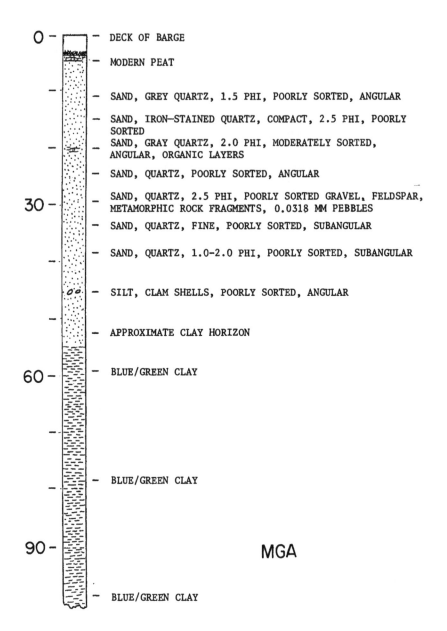

0 — DECK OF BARGE

— MODERN PEAT

— SAND, GREY QUARTZ, 1.5 PHI, POORLY SORTED, ANGULAR

— SAND, IRON-STAINED QUARTZ, COMPACT, 2.5 PHI, POORLY
SORTED

— SAND, GRAY QUARTZ, 2.0 PHI, MODERATELY SORTED,
ANGULAR, ORGANIC LAYERS

— SAND, QUARTZ, POORLY SORTED, ANGULAR

— SAND, QUARTZ, 2.5 PHI, POORLY SORTED GRAVEL, FELDSPAR,
METAMORPHIC ROCK FRAGMENTS, 0.0318 MM PEBBLES

30 —

— SAND, QUARTZ, FINE, POORLY SORTED, SUBANGULAR

— SAND, QUARTZ, 1.0-2.0 PHI, POORLY SORTED, SUBANGULAR

— SILT, CLAM SHELLS, POORLY SORTED, ANGULAR

— APPROXIMATE CLAY HORIZON

60 — — BLUE/GREEN CLAY

— BLUE/GREEN CLAY

90 — MGA

— BLUE/GREEN CLAY

66

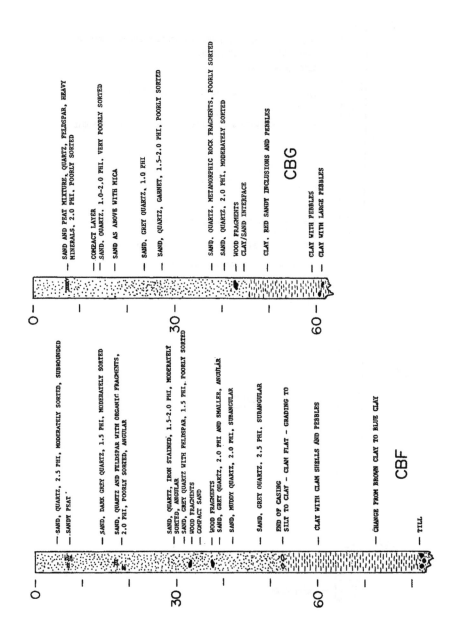

CBG

SAND AND PEAT MIXTURE, QUARTZ, FELDSPAR, HEAVY
MINERALS. 2.0 PHI. POORLY SORTED

COMPACT LAYER
SAND, QUARTZ. 1.0-2.0 PHI. VERY POORLY SORTED

SAND AS ABOVE WITH MICA

SAND, GREY QUARTZ. 1.0 PHI

SAND, QUARTZ, GARNET, 1.5-2.0 PHI, POORLY SORTED

SAND, QUARTZ, METAMORPHIC ROCK FRAGMENTS. POORLY SORTED
SAND, QUARTZ, 2.0 PHI, MODERATELY SORTED
WOOD FRAGMENTS
CLAY/SAND INTERFACE

CLAY, RED SANDY INCLUSIONS AND PEBBLES

CLAY WITH PEBBLES
CLAY WITH LARGE PEBBLES

CBF

SAND, QUARTZ, 2.5 PHI, MODERATELY SORTED, SUBROUNDED
SANDY PEAT

SAND, DARK GREY QUARTZ, 1.5 PHI, MODERATELY SORTED

SAND, QUARTZ AND FELDSPAR WITH ORGANIC FRAGMENTS,
2.0 PHI, POORLY SORTED, ANGULAR

SAND, QUARTZ, IRON STAINED, 1.5-2.0 PHI, MODERATELY
SORTED, ANGULAR
SAND, GREY QUARTZ WITH FELDSPAR, 1.5 PHI, POORLY SORTED
WOOD FRAGMENTS
COMPACT SAND
WOOD FRAGMENTS
SAND, GREY QUARTZ, 2.0 PHI AND SMALLER, ANGULAR

SAND, MUDDY QUARTZ, 2.0 PHI, SUBANGULAR

SAND, GREY QUARTZ. 2.5 PHI. SUBANGULAR

END OF CASING
SILT TO CLAY - CLAM FLAT - GRADING TO

CLAY WITH CLAM SHELLS AND PEBBLES

CHANGE FROM BROWN CLAY TO BLUE CLAY

TILL

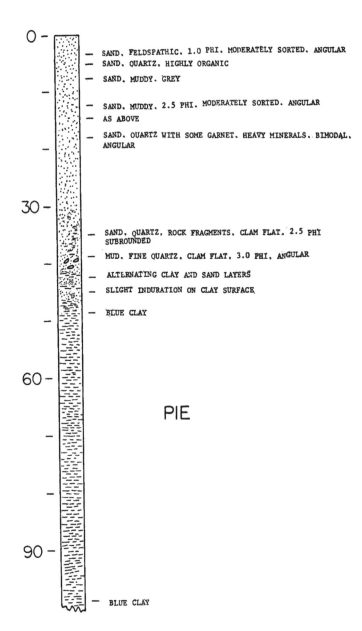

0 —
— SAND, FELDSPATHIC, 1.0 PHI, MODERATELY SORTED, ANGULAR
— SAND, QUARTZ, HIGHLY ORGANIC
— SAND, MUDDY, GREY

— SAND, MUDDY, 2.5 PHI, MODERATELY SORTED, ANGULAR
— AS ABOVE
— SAND, QUARTZ WITH SOME GARNET, HEAVY MINERALS, BIMODAL, ANGULAR

30 —

— SAND, QUARTZ, ROCK FRAGMENTS, CLAM FLAT, 2.5 PHI SUBROUNDED
— MUD, FINE QUARTZ, CLAM FLAT, 3.0 PHI, ANGULAR
— ALTERNATING CLAY AND SAND LAYERS
— SLIGHT INDURATION ON CLAY SURFACE

— BLUE CLAY

60 —

PIE

90 —

— BLUE CLAY

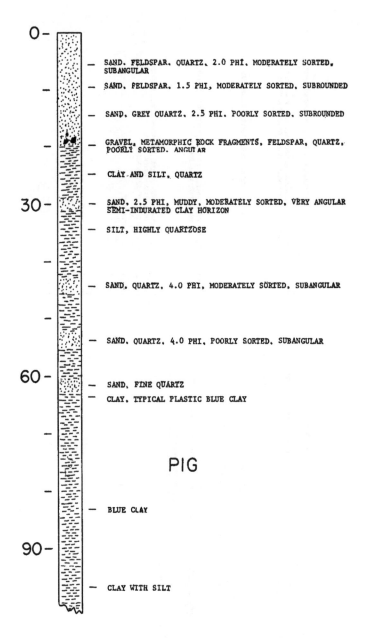

SAND, FELDSPAR, QUARTZ, 2.0 PHI, MODERATELY SORTED, SUBANGULAR

SAND, FELDSPAR, 1.5 PHI, MODERATELY SORTED, SUBROUNDED

SAND, GREY QUARTZ, 2.5 PHI, POORLY SORTED, SUBROUNDED

GRAVEL, METAMORPHIC ROCK FRAGMENTS, FELDSPAR, QUARTZ, POORLY SORTED, ANGULAR

CLAY AND SILT, QUARTZ

SAND, 2.5 PHI, MUDDY, MODERATELY SORTED, VERY ANGULAR SEMI-INDURATED CLAY HORIZON

SILT, HIGHLY QUARTZOSE

SAND, QUARTZ, 4.0 PHI, MODERATELY SORTED, SUBANGULAR

SAND, QUARTZ, 4.0 PHI, POORLY SORTED, SUBANGULAR

SAND, FINE QUARTZ

CLAY, TYPICAL PLASTIC BLUE CLAY

PIG

BLUE CLAY

CLAY WITH SILT

69

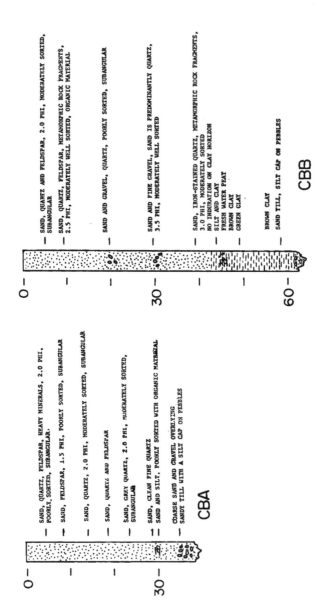

CBA

SAND, QUARTZ, FELDSPAR, HEAVY MINERALS, 2.0 PHI, POORLY SORTED, SUBANGULAR.

SAND, FELDSPAR, 1.5 PHI, POORLY SORTED, SUBANGULAR

SAND, QUARTZ, 2.0 PHI, MODERATELY SORTED, SUBANGULAR

SAND, QUARTZ AND FELDSPAR

SAND, GREY QUARTZ, 2.0 PHI, MODERATELY SORTED, SUBANGULAR

SAND, CLEAN FINE QUARTZ
SAND AND SILT, POORLY SORTED WITH ORGANIC MATERIAL

COARSE SAND AND GRAVEL OVERLYING
SANDY TILL WITH A SILT CAP ON PEBBLES

CBB

SAND, QUARTZ AND FELDSPAR, 2.0 PHI, MODERATELY SORTED, SUBANGULAR

SAND, QUARTZ, FELDSPAR, METAMORPHIC ROCK FRAGMENTS, 2.5 PHI, MODERATELY WELL SORTED, ORGANIC MATERIAL

SAND AND GRAVEL, QUARTZ, POORLY SORTED, SUBANGULAR

SAND AND FINE GRAVEL, SAND IS PREDOMINANTLY QUARTZ, 3.5 PHI, MODERATELY WELL SORTED

SAND, IRON-STAINED QUARTZ, METAMORPHIC ROCK FRAGMENTS, 3.0 PHI, MODERATELY SORTED
NO INDURATION ON CLAY HORIZON
SILT AND CLAY
FRESH WATER PEAT
BROWN CLAY
GREEN CLAY

BROWN CLAY

SAND TILL, SILT CAP ON PEBBLES

70

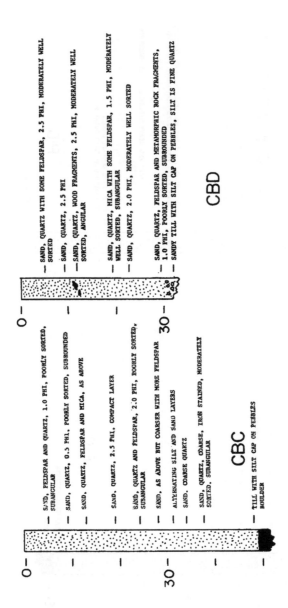

CBD

CBC

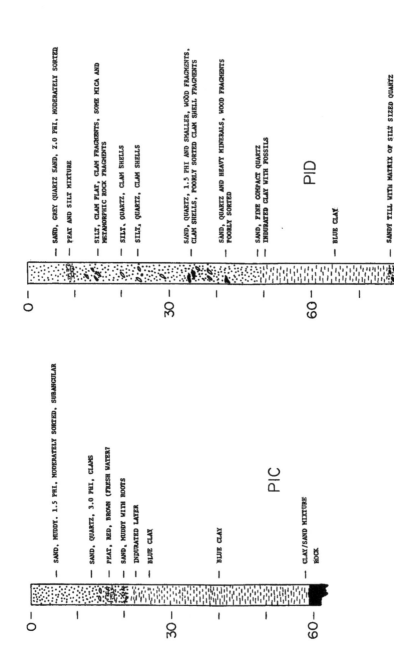

PIC

SAND, MUDDY, 1.5 PHI, MODERATELY SORTED, SUBANGULAR

SAND, QUARTZ, 3.0 PHI, CLAMS

PEAT, RED, BROWN (FRESH WATER?)

SAND, MUDDY WITH ROOTS

INDURATED LAYER

BLUE CLAY

BLUE CLAY

CLAY/SAND MIXTURE

ROCK

PID

SAND, GREY QUARTZ SAND, 2.0 PHI, MODERATELY SORTED

PEAT AND SILT MIXTURE

SILT, CLAM FLAT, CLAM FRAGMENTS, SOME MICA AND METAMORPHIC ROCK FRAGMENTS

SILT, QUARTZ, CLAM SHELLS

SILT, QUARTZ, CLAM SHELLS

SAND, QUARTZ, 1.5 PHI AND SMALLER, WOOD FRAGMENTS, CLAM SHELLS, POORLY SORTED CLAM SHELL FRAGMENTS

SAND, QUARTZ AND HEAVY MINERALS, WOOD FRAGMENTS POORLY SORTED

SAND, FINE COMPACT QUARTZ INDURATED CLAY WITH FOSSILS

BLUE CLAY

SANDY TILL WITH MATRIX OF SILT SIZED QUARTZ

72

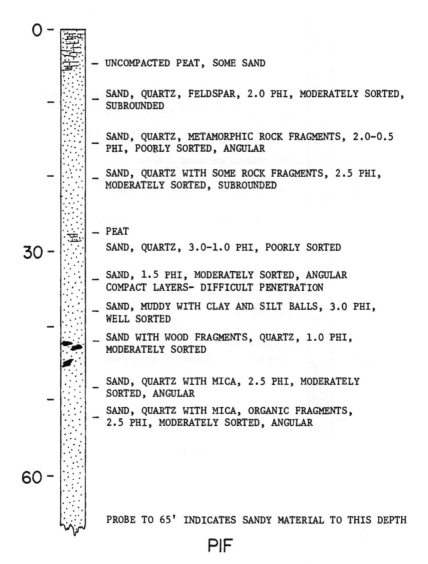

UNCOMPACTED PEAT, SOME SAND

SAND, QUARTZ, FELDSPAR, 2.0 PHI, MODERATELY SORTED, SUBROUNDED

SAND, QUARTZ, METAMORPHIC ROCK FRAGMENTS, 2.0-0.5 PHI, POORLY SORTED, ANGULAR

SAND, QUARTZ WITH SOME ROCK FRAGMENTS, 2.5 PHI, MODERATELY SORTED, SUBROUNDED

PEAT

SAND, QUARTZ, 3.0-1.0 PHI, POORLY SORTED

SAND, 1.5 PHI, MODERATELY SORTED, ANGULAR COMPACT LAYERS- DIFFICULT PENETRATION

SAND, MUDDY WITH CLAY AND SILT BALLS, 3.0 PHI, WELL SORTED

SAND WITH WOOD FRAGMENTS, QUARTZ, 1.0 PHI, MODERATELY SORTED

SAND, QUARTZ WITH MICA, 2.5 PHI, MODERATELY SORTED, ANGULAR

SAND, QUARTZ WITH MICA, ORGANIC FRAGMENTS, 2.5 PHI, MODERATELY SORTED, ANGULAR

PROBE TO 65' INDICATES SANDY MATERIAL TO THIS DEPTH

PIF

73

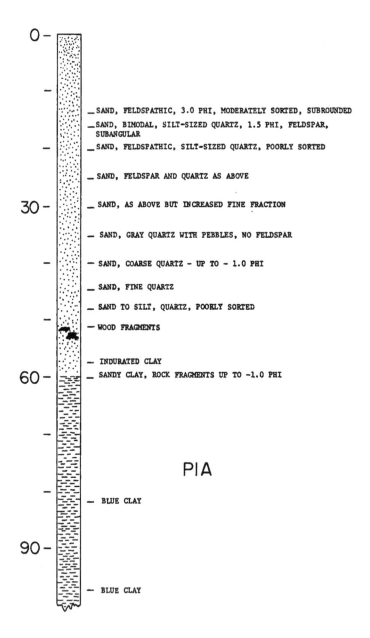

0 —

— SAND, FELDSPATHIC, 3.0 PHI, MODERATELY SORTED, SUBROUNDED
— SAND, BIMODAL, SILT-SIZED QUARTZ, 1.5 PHI, FELDSPAR,
   SUBANGULAR
— SAND, FELDSPATHIC, SILT-SIZED QUARTZ, POORLY SORTED

— SAND, FELDSPAR AND QUARTZ AS ABOVE

30 — — SAND, AS ABOVE BUT INCREASED FINE FRACTION

— SAND, GRAY QUARTZ WITH PEBBLES, NO FELDSPAR

— SAND, COARSE QUARTZ - UP TO - 1.0 PHI

— SAND, FINE QUARTZ

— SAND TO SILT, QUARTZ, POORLY SORTED

— WOOD FRAGMENTS

— INDURATED CLAY
— SANDY CLAY, ROCK FRAGMENTS UP TO -1.0 PHI
60 —

PIA

— BLUE CLAY

90 —

— BLUE CLAY

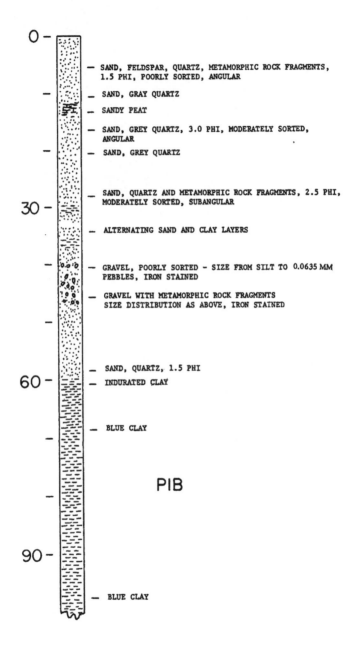

0 —

— SAND, FELDSPAR, QUARTZ, METAMORPHIC ROCK FRAGMENTS,
  1.5 PHI, POORLY SORTED, ANGULAR

— SAND, GRAY QUARTZ

— SANDY PEAT

— SAND, GREY QUARTZ, 3.0 PHI, MODERATELY SORTED,
  ANGULAR

— SAND, GREY QUARTZ

— SAND, QUARTZ AND METAMORPHIC ROCK FRAGMENTS, 2.5 PHI,
  MODERATELY SORTED, SUBANGULAR

30 —

— ALTERNATING SAND AND CLAY LAYERS

— GRAVEL, POORLY SORTED - SIZE FROM SILT TO 0.0635 MM
  PEBBLES, IRON STAINED

— GRAVEL WITH METAMORPHIC ROCK FRAGMENTS
  SIZE DISTRIBUTION AS ABOVE, IRON STAINED

— SAND, QUARTZ, 1.5 PHI

60 — — INDURATED CLAY

— BLUE CLAY

PIB

90 —

— BLUE CLAY

75

DOCUMENT CONTROL DATA - R & D

(Security classification of title, body of abstract and indexing annotation must be entered when the overall report is classified)

1. ORIGINATING ACTIVITY *(Corporate author)*	2a. REPORT SECURITY CLASSIFICATION
Department of the Army Coastal Engineering Research Center Kingman Building Fort Belvoir, Virginia 22060	UNCLASSIFIED
	2b. GROUP

3. REPORT TITLE

PLEISTOCENE-HOLOCENE SEDIMENTS INTERPRETED BY SEISMIC REFRACTION AND WASH-BORE SAMPLING, PLUM ISLAND-CASTLE NECK, MASSACHUSETTS.

4. DESCRIPTIVE NOTES *(Type of report and inclusive dates)*

5. AUTHOR(S) *(First name, middle initial, last name)*

Eugene G. Rhodes

6. REPORT DATE	7a. TOTAL NO. OF PAGES	7b. NO. OF REFS
July 1973	75	31

8a. CONTRACT OR GRANT NO.	9a. ORIGINATOR'S REPORT NUMBER(S)
b. PROJECT NO.	Technical Memorandum No. 40
c.	9b. OTHER REPORT NO(S) *(Any other numbers that may be assigned this report)*
d.	

10. DISTRIBUTION STATEMENT

Approved for public release; distribution unlimited.

11. SUPPLEMENTARY NOTES	12. SPONSORING MILITARY ACTIVITY
	Department of the Army Coastal Engineering Research Center Kingman Building Fort Belvoir, Virginia 22060

13. ABSTRACT

The wash-bore method of soil sampling was found to be an excellent technique for subsurface study in coastal areas. Phenomena to be considered when interpreting seismic refraction records include a) the "blind zone," b) the non-zero time intercept, c), time gaps in the time-distance photo over buried peat, and d) variable thicknesses of dry sand layers.

The seismic method successfully located Pleistocene and bedrock topography. However, glaciomarine clay did not show a seismic contrast with respect to sandy, water-soaked sediments. Topography exposed during lower sea level has a dominant influence on modern coastal geology. Barrier islands became anchored on Pleistocene features as the sea level rose and deposition occurred in the estuaries behind the barrier beaches. Major channels of the estuaries migrated landward with the sea-level rise.

No radiometric dates were determined from study samples but the sedimentary stratigraphy fits the time frame of other investigators.

DD ,FORM,.**1473** REPLACES DD FORM 1473, 1 JAN 64, WHICH IS OBSOLETE FOR ARMY USE.

Rhodes, Eugene G.
  Pleistocene-holocene sediments interpreted by seismic refraction
and wash-bore sampling, Plum Island-Castle Neck, Massachusetts.
Fort Belvoir, Va., U.S. Coastal Engineering Research Center, 1973.
75 p. illus. (U.S. Coastal Engineering Research Center, Technical
Memorandum No. 40); (Contract DACW 72-70-C-0029).

  Thesis (M.S.) - University of Massachusetts.
  Bibliography: p. 59-61.
  The wash-bore method of soil sampling was found to be an excellent
means for subsurface study in coastal areas. Considerations in
interpretation of seismic refraction records are a) the blind z n 9
b) the non-zero time intercept, c) time gaps in the time-distance
plots over buried peat, and d) variable thicknesses of dry sand
layers. The seismic method successfully located buried Pleistocene
and bedrock topography.

  1. Geomorphology - Plum Island, Mass. 2. Stratigraphic geology-
Pleistocene. 3. Seismic refraction - Plum Island, Mass. I. Title.
(Series)(Contract)

TC203   .U58ltm   no. 40   627   .U58ltm no. 40

Rho de, 1 ene G.
  Pleistocene-holocene sediments interpreted by seismic refract:
and wash-bore sampling, Plum Island-Castle N ko Massachusetts.
F rt Belvoir, Va., U.S. Coastal Engineering Research Center, 19:
75 p. illus. (U.S. Coastal Engineering Research Center, Techni
Memorandum No. 10);   ract DCW 72-70-C- 0g.

  Th sis (M.S.) - University of Massachusetts.
  Bibliography: p. 59-61.
  The wash-bore method of soil sampling was found to be an exce:
interpretation of seismic refraction records are a) the blind z:
b) the non-zero time intercept, c) time gaps in the time-distan:
plots over buried peat, and d) variable thicknesses of dry sand
layers. The seismic method successfully located buried Pleistoc
and bedrock topography.

  1. Geomorphology - Plum Island, Mass. 2. Stratigraphic geo:
Pleistocene. 3. Seismic refraction - Plum Island, Mass. I. Tit
(Series)(Contract)

TC203   .U58ltm   no. 40   627   .U58ltm n:

Rhodes, Eugene G.
  Pleistocene-holocene sediments interpreted by seismic refract:
and wash-bore sampling, Plum Island-Castle Neck, Massachusetts.
F rt Belvoir, Va., U.S. Coastal Engineering Research Center, 19:
75 p. illus. (U.S. Coastal Engineering Research Center, Techn:
Memorandum No. 40); (Contract DACW 72-70-C-0029).

  Thesis (M.S.) - University of Massachusetts.
  Bibliography: p. 59-61.
  The wash-b re method of soil sampling was found to be an exce:
m ans for subsurface study in coastal areas. Considerations in
interpretation of seismic refraction records are a) the blind z:
b) the non-zero time intercept, c) time gaps in the time-distan:
plots over buried peat, and d) variable thicknesses of dry sand
layers. The seismic m th d successfully located buried Pleisto:
and bedrock topography.

  1. Geomorphology - Plum Island, Mass. 2. Stratigraphic geo:
Pleistocene. 3. Seismic refraction - Plum Island, Mass. I. Tit
(Series)(Contract)

TC203   .U58ltm   no. 40   627   .U58ltm n:

es, 1 ene G.
  Pleistocene-holocene sediments interpreted by seismic refraction
and wash-bore sampling, Plum Island-Castle Neck, Massachusetts.
Fort Belvoir, Va., U.S. Coastal Engineering Research Center, 1973.
75 p. illus. (U.S. Coastal Engineering Research Center, Technical
Memorandum No. 40); (Contract DACW 72-70-C-0029).

  Thesis (M.S.) - University of Massachusetts.
  Bibliography: p. 59-61.
  The wash-b re method of s id sampling was found to be an excellent
means for subsurface study in coastal areas. Considerations in
interpretation of s iemic refraction records are a) the blind z n 9
b) the non-zero ime intercept, c) time gaps in the time-distance
plots over buried p atg and d) variable thicknesses of dry sand
layers. The seismic method successfully located buried Pleistocene
and bedrock topography.

  1. Geomorphology - Plum Island, Mass. 2. Stratigraphic geology-
Pleistocene. 3. Seismic refraction - Plum Island, Mass. I. Title.
(Series)(Contract)

TC203   .U58ltm   no. 40   627   .U58ltm no. 40

Rhodes, Eugene G.
  Pleistocene-holocene sediments interpreted by seismic refraction
and wash-bore sampling, Plum Island-Castle Neck, Massachusetts.
Fort Belvoir, Va., U.S. Coastal Engineering Research Center, 1973.
  75 p. illus. (U.S. Coastal Engineering Research Center, Technical
Memorandum No. 40); (Contract DACW 72-70-C-0029).

  Thesis (M.S.) - University of Massachusetts.
  Bibliography: p. 59-61.

  The wash-bore method of soil sampling was found to be an excellent
means for subsurface study in coastal areas. Considerations in
interpretation of seismic refraction records are a) the blind zone,
b) the non-zero time intercept, c) time gaps in the time-distance
plots over buried p atq and d) variable thicknesses of dry sand
layers. The seismic method successfully located buried Pleistocene
and bedrock topography.

  1. Geomorphology - Plum Island, Mass. 2. Stratigraphic geology-
Pleistocene. 3. Seismic refraction - Plum Island, Mass. I. Title.
(Series)(Contract)
TC203   .US81tm   no. 40   627   .US81tm no. 40

Rhodes, Eugene G.
  Pleistocene-holocene sediments interpreted by seismic refract:
and wash-bore sampling, Plum Island-Castle Neck, Massachusetts.
F rt Belvoir, Va., U.S. Coastal Engineering Research Center, 19
Memorandum No. 40); (Contract DACW 72-70-C-0029).

  Thesis (M.S.) - University of Massachusetts.
  Bibliography: p. 59-61.

  The wash-bore method of s id sampling was found to be an x ce
means for subsurface study in coastal areas. Considerations in
interpretation of seismic refraction records are a) the blind z
b) the non-zero time intercept, c) time gaps in the time-distan
plots over buried p atq and d) variable thicknesses of dry sand
layers. The seismic method successfully located buried Pleistoc
and bedrock topography.

  1. Geomorphology - Plum Island, Mass. 2. Stratigraphic geo
Pleistocene. 3. Seismic refraction - Plum Island, Mass. I. Tit
(Series)(Contract)
TC203   .US81tm   no. 40   627   .US81tm n

Rhodes, Eugene G.
  Pleistocene-holocene sediments interpreted by seismic refraction
and wash-bore sampling, Plum Island-Castle Neck, Massachusetts.
Fort Belvoir, Va., U.S. Coastal Engineering Research Center, 1973.
  75 p. illus. (U.S. Coastal Engineering Research Center, Technical
Memorandum No. 40); (Contract DACW 72-70-C-0029).

  Thesis (M.S.) - University of Massachusetts.
  Bibliography: p. 59-61.

  The wash-bore method of soil sampling was found to be an excellent
means for subsurface study in coastal areas. Considerations in
interpretation of seismic refraction records are a) the blind zone,
b) the non-zero time intercept, c) time gaps in the time-distance
plots over buried p atq and d) variable thicknesses of dry sand
layers. The seismic method successfully located buried Pleistocene
and bedrock topography.

  1. Geomorphology - Plum Island, Mass. 2. Stratigraphic geology-
Pleistocene. 3. Seismic refraction - Plum Island, Mass. I. Title.
(Series)(Contract)
TC203   .US81tm   no. 40   627   .US81tm no. 40

Rhodes, Eugene G.
  Pleistocene-holo ene sediments interpreted by seismic refract:
and wash-bore sampling, Plum Island-Castle Neck, Massachusetts.
F rt Belvoir, Va., U.S. Coastal Engineering Research Center, 19
Memorandum No. 40); (Contract DACW 72-70-C-0029).

  Thesis (M.S.) - University of Massachusetts.
  Bibliography: p. 59-61.

  The wash-bore method of soil sampling was found to be an ex o
means for subsurface study in c astal areas. Considerations in
interpretation of seismic refraction records are a) the blind z
b) the non-zero time intercept, c) time gaps in the time-distan
plots over buried p atq and d) variable thicknesses of dry sand
layers. The seismic method successfully located buried Pleistoc
and bedrock topography.

  1. Geomorphology - Plum Islan d Mass. 2. Stratigraphic geo
Pleistocene. 3. Seismic refraction - Plum Island, Mass. I. Tit
(Series)(Contract)
TC203   .US81tm   no. 40   627   .US81tm n

Lightning Source UK Ltd.
Milton Keynes UK
UKHW010735131218
333917UK00009B/849/P